A Priori Revisability in Science

A Priori Revisability in Science

By

Boris D. Grozdanoff

CAMBRIDGE
SCHOLARS
PUBLISHING

A Priori Revisability in Science, by Boris D. Grozdanoff

This book first published 2014

Cambridge Scholars Publishing

12 Back Chapman Street, Newcastle upon Tyne, NE6 2XX, UK

British Library Cataloguing in Publication Data
A catalogue record for this book is available from the British Library

ISBN (10): 1-4438-5822-6, ISBN (13): 978-1-4438-5822-9

TO MY NAIA

TABLE OF CONTENTS

FOREWORD

The most influential rationalist model of scientific knowledge is the three-layered model formulated recently by Michael Friedman. At its surface are the empirical laws of nature, such as Newtonian law of gravitation or Einstein's equations for gravitational field. At its deeper second level are the fundamental principles of science that determine the general spatiotemporal framework which enables the formulation and the testing of the empirical laws. At the third level are the philosophical meta-paradigms which guide the transition between scientific paradigms. The central epistemic claim of the model concerns the character of the fundamental principles; according to Friedman they are a priori, that is, they are independent from experience. Yet he is explicit that the principles change under empirical pressure. Friedman's position, however, faces the modern empiricist challenge instead of evading it: he has to explain *how* the principles could still be a priori if they change under empirical pressure. I take his defence, appealing to the old Reichenbachian notion of the constitutive a priori, as inconclusive. My general aim is to put forward a rationalist account of the scientific knowledge and I will attempt a defence of the main epistemic claims in Friedman's model, which I find vulnerable against the prevailing empiricism. The present text provides a contemporary account of the epistemic character of the principles addressing the recent work on the a priori. I argue that at least some principles within natural science are not empirically but a priori revisable, and in this way I respond to the empiricist challenge. In order to build the defence I formulate a general notion of epistemic revisability and I extract from it two corresponding kinds of specific revisabilities: a traditional empirical one and the suggested a priori revisability within natural science. I argue that the latter kind is as vital as the former and that it is also capable of meeting the argument from empirical revisability by providing an epistemic alternative of it. In this way, if some second level principles are shown to evolve through a priori revisions the leading empiricist argument fails. To demonstrate this I analyze two case studies, one from history of geometry and one from history of physics, and I show that the revisions were epistemically a priori and not empirical. The result is a two-fold one. First, a genuine alternative of empirical revisability is developed, and not just for a priori domains like mathematics but for the natural sciences. Second, a new mechanism for the dynamics of science is suggested, namely that scientific knowledge sometimes evolves through empirically independent moves. At the end, these enable a modern epistemic defence of the priori character of the second level principles in Friedman's model and thus help to keep its vitality.

ACKNOWLEDGEMENTS

The present text was written while I was a Marie Curie Fellow at the University of Oxford (2008 – 2010). I am indebted to many people but several deserve a special acknowledgement. For invaluable help both on the notion of a priori and the epistemic problem of scientific thought experiments I greatly need to thank Prof. Nenad Miscevic from the Central European University and Prof. James Robert Brown from the University of Toronto. For illuminating discussions in philosophy of science and philosophy of physics I am indebted to Prof. John Norton from the University of Pittsburgh. For a great help mostly in the form of seminars and lectures I want to thank my supervisor Dr. Oliver Pooley, Prof. Harvey Brown and Prof. Simon Saunders. For assistance on various text related problems I want to thank Assoc. Prof. Lilia Gurova, Prof. Gerald Massey, Prof. Ken Manders and Assoc. Prof. Jennifer Nagel. Most of all, however, I would like to thank my mentor Nenad Miscevic without whose dedicated, intensive and friendly help in virtually every related matter this text would have not been written. It goes without saying that all deficiencies, such as mistakes, omissions, weak or simply wrong arguments are entirely my own responsibility. I am grateful to the group of Philosophy of Physics at the University of Oxford which kindly hosted me for two years and also, I am grateful to the philosophy department at the CEU for all academic support during the last years. At the end, I would like to thank Prof. Hanoch Ben-Yami for a useful criticism.

INTRODUCTION

The change of scientific theories has been in the focus of the philosophy of science ever since the latter's emergence as a philosophical discipline. The founding fathers of the discipline, the logical positivists and most notably Rudolf Carnap[1] and Hans Reichenbach,[2] demonstrated the importance of the problem in a series of influential books and papers. They developed an approach that aimed to accommodate the obvious role experience had in the development of scientific knowledge. Yet, they tried to frame this role in a framework that had rationalistic fundamentals. The clearest illustration provides Reichenbach's early work "Theory of Relativity and A priori Knowledge", where he put forward the notion of constitutive a priori principles which underlie the foundations of scientific knowledge. Later on, among others, Karl Popper[3] and Imre Lakatos[4] took on the problem of scientific change and proposed models for the theoretical dynamics that addressed its epistemic aspect. But perhaps the most influential model in our recent history is the famous model suggested in the early 1960s by the historian of science Thomas Kuhn.[5] He argued that scientific knowledge evolves by following a chain of stable periods (called "scientific paradigms") with clear rules and standards, and revolutionary periods that substitute the previous paradigms with new ones. The central point in this model was the mechanism of the very transition. Kuhn claimed that there is an inherent incommensurability between the different paradigms, and thus he raised the question about the rationality that governs the progress of science with a new force. The model of scientific revolutions explicates the *dynamics* of scientific theories but lacks the *epistemic perspective* of the logical positivistic models. Thus, a tension formed in scientific epistemology.

This tension has been addressed only very recently, on the verge of the new millennium, by Michael Friedman. He proposed a model of scientific knowledge which emerges out of the synthesis between the foundational epistemic principles of logical positivism (and especially the early Reichenbach) and the dynamics of the Kuhnian structure of scientific revolutions. The model is distinct for the epistemic nature of its fundamental principles which, according to Friedman, are independent from experience and are thus a priori. Also, it is notably the most influential contemporary rationalistic model of scientific knowledge. The model, however, meets strong resistance from the predominant scientific epistemology today, the modern empiricism. The main target for criticism in the model is not its dynamics but its epistemic commitment. Scientific empiricists argue against the a priori character of the fundamental principles of science, and thus endanger its alleged epistemic nature. In addition, the criticism bears on the dynamics of the model for if the fundamental principles are not a priori then the whole dynamics of science transforms into empirical dynamics. Therefore, it is important for Friedman to block the epistemic criticism and in particular, this should be done within the context of the modern debates on the a priori. In the exposition and the argumentation of the model, however, this is not what happens. Friedman certainly addresses the issue of empirical fallibility of the alleged a priori principles and yet his approach is not epistemic but semantic. He argues that the principles are a priori and at the same time revisable but his reasoning is that they are not empirically revisable only because they are constitutive for the possibility of their own empirical revisions. This is at the core the Reichenbachian notion of the constitutive a priori, which is nonetheless distant from the centre of the modern debates in epistemology. The property of *being constitutive for* is not difficult to recognize as a semantic and not as an epistemic property, and that is why an epistemic defence of the a priori nature of the principles in the model is *de facto* lacking. It is, nevertheless, more or less universally agreed today, that the fundamental semantic, modal and epistemic distinctions are independent. Therefore, a modern attempt to defend an epistemic notion should be based on epistemic arguments even if semantic considerations are instructive. Such a modern epistemic attempt is, however, not available and this creates a problem for the central epistemic claim in the model.

My main goal in this text is to provide at least a partial solution for this problem. In particular, I will present an epistemic defence for a specific dynamics of the a priori principles in Friedman's model. In order for the defence to be adequate to the modern epistemology I will address and sometimes follow several recent works on the nature of the a priori notion, such as the works of Albert Casullo[6] and Joshua Thurow.[7] If successful, the defence would help both the epistemic nature of the model and its dynamics. In this sense, the project is of significance for current philosophy of science for it would re-establish the best available model of scientific rationalism as an epistemically vital one. In addition, the defence would help to fill in what I take to be an epistemic gap in the model. Namely, Friedman does not answer the problem of fallibility, and consequently, he does not address the change of the fundamental principles epistemically but only semantically. He defends against standard criticism from the empirical fallibility of those principles by an appeal to the old Reichenbachian notion of constitutive a priori which, I argue, is at its core of semantic and not of epistemic nature. Thus, there exists a vacuum for a modern epistemic account of the problem of the dynamics of fundamental principles that argues about their epistemic character purely epistemically and not semantically or modally. For that purpose I will beforehand suggest and develop the epistemic conception of a priori revisability and I will apply it to some fundamental principles in the model. The notion would be developed both positively and negatively. I will positively define its characteristics and I will provide arguments for its epistemic legitimacy. I will negatively define the domain of its validity through narrowing the domain of validity of traditional rival empirical revisability and through criticizing the main empiricist arguments. At the end, I will argue about an actual historical role and significance of the notion in both mathematics and natural science.

The problem of epistemic revisability within natural science did not receive particular attention in current epistemology and has not been explained since Kuhn in sufficient detail and in its full scope. It is predominantly accepted that as far as the epistemic aspect of both general and scientific dynamics is concerned it is naturally of empirical nature. Given the fact of the existence of two epistemic kinds, a posteriori and a priori, this results thereafter in an imbalance in symmetry. Therefore, I go on and explore the notion of symmetric epistemic dynamics. Its suggested role in cases like the epistemic problems of Friedman's model of scientific knowledge, demonstrates that this area in epistemology has been unjustly neglected and deserves much greater consideration. Both notions that I suggest in the text, the general notion of *epistemic revisability* and the specific notion of *a priori revisability*, are meant as provoking a more intensive discussion. If the main line of argumentation in the text is correct, then the predominant view that the empirical revisability is the sole kind of epistemic revisability and that the dynamics of scientific knowledge is only empirical dynamics should be reconsidered.

The standard empiricist argument against the a priori steps on fallibility: a proposition could not be a priori if it is empirically revisable or worse, empirically revised. I attempt to show that the argument fails and in this I follow two main lines of argumentation. The first one is epistemic and argues that the argument rests on a tacit assumption about the relationships between the epistemic kinds of justification and revision which turns out to be mistaken. Thus the argument could not claim that the epistemic kind of a revision could actually bear on the epistemic kind of the justification. Further, since the empiricist could still insist that, even if the epistemic kind of a revision does not bear on the epistemic kind of the proposition, it nevertheless contributes to the *overall* dependence of the proposition on experience; and this hits directly into the nature of the a priori as being independent from experience. Here I argue that in fact alternative epistemic revisions which are non-empirical are, first, conceivable and second, historically actual. I develop the conception of a priori revisability, which opens the door for a priori justified propositions to be independent from experience, in response to the second step of the empiricist argument. I argue that some a priori justified propositions are a priori revisable and that thus they are independent from experience in a two-fold way; I take this to be sufficient to establish their overall epistemic status as a priori propositions. The second line of argumentation is the historical one. My illustrations are both from the domains of mathematics and natural science. I propose two case studies of what I take to be influential revisions in history of mathematics and science, and I argue that they are revisions of both a priori held and a priori revised propositions. In this way I add flesh to the conceived of epistemic conception of a priori revisability. The whole project is oriented towards the concrete goal to seal Friedman's influential neo-Kantian model of scientific knowledge against the standard empiricist criticism and for that purpose I assume the general framework of Friedman's model.

The line of argumentation in the text has the following structure. The suggested conception of a priori revisability starts with the uncontroversially recognized notion of empirical revisability and moves towards the more abstract notion of epistemic revisability. From here, given the availability of the two epistemic kinds, the empirical and the a priori kind, it follows the natural step of conceiving of the epistemically complementary sort of revision: the a priori revision within natural science. After establishing it as a legitimate epistemic alternative of empirical revisability within natural science, I argue that it is also actually functioning by providing illustrations specifically taken from the second level of Friedman's model of scientific knowledge, namely the level of fundamental a priori principles. Further, I suggest an epistemic analysis of two influential cases of revision in geometry (from the domain of mathematics) and physics (from the domain of natural science), both complying with the requirement for the second level principles which Friedman takes "to define the fundamental spatiotemporal framework within which alone the formulation and empirical testing of base principles is possible". I argue that the revision of the famous 5th postulate of the Euclidean geometry and the revision of the principle of absolute simultaneity in physics are a priori revisions of a priori principles. For the first, more obvious case, I follow modern discussions in philosophy of mathematics (and especially ones that step on semantics) as well as the actual historical process of the revision. For the second case I look at thought experiments and I employ a recent interpretation of their role in natural science, put forward by James Robert Brown, which argues that some thought experiments lead to a priori knowledge about the world. In the spirit of Kuhn, who famously claimed that thought experiments often happen at the verge of the shift between scientific paradigms, I argue that sometimes, rare as it might be, science progresses by a priori moves. And in particular, these moves often concern the formulation and the revision of precisely what Michael Friedman calls "a priori constitutive principles". In order to defend this conception and my own analysis I respond to the influential empiricist account on the epistemic role of thought experiments in science, the logical reconstructability account by John Norton. In this way the a priori revisability account presented here coordinates the neo-Kantian account of scientific knowledge, proposed by Friedman, the platonic account of thought experiments of Brown and the Kuhnian dynamics of scientific paradigms. It by no means eliminates all incompatibilities between those, which in some cases, like the case of the underlying metaphysics, seem to me too difficult to overcome. Yet, it points to an important mechanism, the mechanism of a priori revisability within natural science, that is capable of regulating the common epistemic compatibility among them. And at the end of the day it is this compatibility that is of significant importance for the rationalist project.

Notes

[1] Carnap, R. [1928] *The logical Structure of the world*. Berkeley and Los Angeles: University of California Press, 1967, and Carnap, R. [1934] *The Logical Syntax of Language*, London: Kegan Paul, 1937. Also Carnap, R. [1936] "Testability and Meaning" and Carnap, R. [1947] *Meaning and Necessity*: A Study in Semantics and Logic, Chicago: University of Chicago Press.

[2] Reichenbach, Hans [1965] *Theory of Relativity and A priori Knowlegde*, Los Angeles, University of California Press.

[3] Popper, K. [1959] *The Logic of Scientific Discovery*. Hutchinson, London, 1959. and Popper, K. [1963] *Conjectures and Refutations: The Growth of Scientific Knowledge*. Routledge, London.

[4] Lakatos [1978] *The Methodology of Scientific Research Programmes: Philosophical Papers Volume 1*. Cambridge: Cambridge University Press.

[5] Kuhn, T. [1962] *The Structure of Scientific Revolutions*, 1st. ed., Chicago: Univ. of Chicago Press.

[6] Casullo, Albert [1988] "Revisability, Reliabilism and A priori knowledge" in *Philosophy and Phenomenological Research*, 49, pp. 187 – 213, and Casullo, Albert [2003] *A priori Justification*, Oxford University Press, NY.

[7] Thurow, J. [2006] "Experientially Defeasible A priori justification' in *The Philosophical Quarterly*, Vol. 56, No. 225, pp. 596 – 602.

CHAPTER I

FRIEDMAN'S MODEL OF SCIENTIFIC KNOWLEDGE

The chapter introduces the philosophical context within which arises the model of scientific knowledge put forward recently by Michael Friedman. The model attempts to respond to the problems of empiricism by suggesting a mechanism for the formation and development of scientific knowledge. I argue that the model is vulnerable to contemporary criticism from empirical revisability. Also, I argue that Friedman's own defense against this criticism fails because it borrows an outdated formulation of the notion of a priori (the Reichenbachian constitutive a priori) instead of developing a modern one. I suggest that, in order to keep the a priori epistemic character of the principles in the model, Friedman needs a stronger notion of the a priori, which is capable of meeting the empiricist criticism. This notion should be developed within the context of the current debates on the nature of the a priori, and should be able to accommodate the revisability kind of independence from experience. Section one introduces the philosophical background against which Friedman's model is built. Section two presents the three-layered structure of the model and discusses the problem of epistemic dynamics. Section three examines the model with respect to the problem of empirical revisability. Section four presents and critically explores the Reichenbachian notion of constitutive a priori which Friedman accommodates in the model. Section five suggests the new, epistemically stronger account of a priori revisability as a solution to the epistemic problem of the model.

The Epistemic Dynamics of Scientific Theories

The philosophical debate over the nature of a priori has been traditionally an epistemic debate. Epistemic problems in philosophy of science, however, need more intimate links with the results of pure epistemology for, as a rule, they often just touch upon the surface of the latter. Notable exceptions nevertheless exist, such as Reichenbach's *The Theory of Relativity and A priori Knowledge*, where he unites the depth of scientific analysis with the precision of pure epistemology. Other philosophers of science do equally well; Rudolph Carnap and Willard Quine provide further illustrations. Yet, it was not only until recently that such expert epistemologists again set the tone in the widely autonomous field of philosophy of science; Michael Friedman's model being a central theme. He follows closely in the steps of Reichenbach and Carnap and develops a modern setting for a reinterpretation of some of their main ideas; the concept of relativized a priori is the prominent one among them. Friedman is perhaps the most influential rationalist in present-day philosophy of science, a field currently resting primarily on empiricist grounds. Epistemic issues are thus of even greater importance for him, for he has to build an a priori island within a sea of empirical dependencies. Deficiencies in his position would be therefore much less tolerated than in pure epistemology, where the stakes are a bit more technical and not so engaging as in science. The present text is directed towards removing what I take to be an epistemic deficiency in the heart of Friedman's scientific model and addresses its epistemic framework. Let me provide some background.

A major theme in twentieth century philosophy of science has been the change of scientific theories. A significant amount of literature has been devoted to clarifying this problem, starting with the logical empiricists and continuing at present day with modern empiricist accounts. Among the most influential accounts is the famous model of scientific theories change proposed by Thomas Kuhn in 1962.[1] In his model, science develops through a sequence of stable periods and revisionary revolutions, and theories from one stage of development are incommensurable with theories from another stage. Kuhn's model created quite a stir when it first appeared and continues to be in the center of the contemporary discussion. Besides the rich controversy around the incommensurability thesis, it is the very nature of the scientific dynamics that is at the core of the model. A large portion of its influence is due to the proposed explanation of its nature; the dynamics of science has been in the main focus of interest even more after *The Structure of Scientific Revolutions* came out. The underlying epistemic ground for virtually all major available positions that address the problem has been one or another form of contemporary empiricism. The empiricist explanation of the problem of scientific dynamics ties it with scientific experience: it is only through experience that we could know if our hypotheses and theories are correct or on the right track, and it is only through experience that we could know if their modifications or substitutes are correct too. Together with the source epistemic claim that most if not all of our knowledge comes from experience, this furnishes an epistemically complete model of scientific change. Perhaps it would not be much of an exaggeration to say that this model is practically dominant in the contemporary scientific epistemology.

Within it scientific dynamics is empirical dynamics and any deviation from this view is as a rule regarded with a healthy dose of skepticism. The empirical dynamics is nevertheless far from uncontroversial. The increasing detachment of theoretical entities of contemporary physics from our immediate experience provides a useful illustration. The essential role of disciplines traditionally considered as non-empirical (such as mathematics and logic) for the scientific enterprise provides another illustration. The nature of the discovered micro-physical world has revealed, perhaps unexpectedly, another problem: there are components of physical reality, well formed within our best theories, that we could not possibly have (sufficiently rich) information about through measurement and experience. And this seems not to be due to imperfections of our experiential apparatus but due to the very nature of the way the things are, as revealed by the theories. A paradox emerges: on the one hand empiricism is the best weapon of contemporary science, and on the other it seems that the same science imposes limits to its application which do not look easily surmountable. If science is to avoid a fall into skepticism it has to address this problem.

One way of addressing it is to attribute suspicion to the theories and argue that because of their deficiencies we might be regarding as empirically unsolvable problems that are actually empirically solvable, and we just have to wait until technology progresses sufficiently so that we become able to probe the problematic domains. Promising as it sounds, this approach has a difficult task; for even if our theories are imperfect, as they certainly are, they are probably not so imperfect as to err on all things we could not have experience for. For example, it does not seem very likely for any empiricist theory of knowledge based on standard relativistic causality to be able to predict with sufficient certainty anything within causally decoherent domains. If I am an empiricist in a light-cone A and I want to know something happening within a light-cone B but A and B do not have common points, then any direct causal signal that could possibly carry some information from A to B or B to A would violate the physical principle of the speed of light.[2] For prediction would need a possible causal connection with the domain of the prediction and, Special Relativity embraced, no such connection could be claimed to exist without violating the limiting principle of light, one formulation of which is that no signal whatsoever could travel faster than the speed of light.

Further, within quantum mechanics there are limits to what we can know via measurement or observation, not because of imperfect technology, but because of the way the world is. Thus, we could not know simultaneously both the precise values of the position and the momentum of an elementary particle: the more we know about its position the less we know about its momentum and vice versa. The regulating principle behind this intuition-defying phenomenon is the quantum mechanical uncertainty principle and again, it seems that this limitation on our knowledge is imposed by reality itself and not by imperfections in our equipment or our theories. To give a last illustration, mathematics and logic are essential parts of contemporary science, and yet it is far from clear what is the nature of the relation between pure mathematics and logic, on the one hand, and physical world, on the other. Refusing to acknowledge the existence of a problem here is of no help. The domains of pure and applied mathematics are quite distinct, and the coordination between mathematical sub-disciplines and physical events have been notoriously difficult to spell out convincingly. Semantic considerations worsen the problem: if purely mathematical propositions and physical propositions are about to have the same truth conditions, and because the prevailing semantics in current science is one or another formulation of Tarskian semantics,[3] we would end up having a platonic account about mathematics, and usually this is not an easy pill to swallow for a natural scientist. If, on the other hand, they are not to have the same semantics, then it is not at all obvious why this should be the case; how much confidence we would be left with in our theories of knowledge, if semantics is non-homogenous even within the domain of science (as far as purely mathematical propositions and logical propositions are in a way a subset of the set of scientific propositions).

Another way of addressing the problem it is to deny that empiricism is giving us the whole story. If not all of our knowledge comes from experience then the skeptical problem loses most of its strength since, at least in principle, we could have gained some knowledge in a non-empirical way; and there are good prospects to expect that the limits imposed on observation and measurement would not limit the alleged non-empirical ways of receiving information, whatever they may turn out to be. In the case of scientific knowledge, we have to distinguish between two different senses of denying the empiricist thesis. The first sense is to deny the source claim that all knowledge comes from experience. The second sense is to deny that *experience is the driving engine behind scientific dynamics*. None of the senses would be sufficient by itself to overturn the empiricist model of scientific change. Even if some of our knowledge turns out not to come from experience, the experience might still serve its regulatory function of driving scientific changes through confirming or disconfirming scientific hypotheses and theories. And even if there is an alternative non-empirical engine behind the dynamics, the very hypotheses and theories might still well come from experience. This demonstrates the complexity of the task before the non-empiricist: on the one hand, she has to draw philosophically and scientifically respectful conceptions of non-empirical justification and revision, and on the other hand, she has to show that these conceptions have actual counterparts in historically interesting cases in science. Even a superficial acquaintance with contemporary scientific epistemology would show how unlikely to be resolved this task is. And yet there are good reasons to acknowledge the problems before scientific empiricism and even better ones to attempt to come up with an epistemically improved solution. An important qualification is in place here. None of the rival epistemic models would ever purport to substitute in full the empiricist model. The goal is not to present a model which is so radically different from the empiricist one that claims that all our knowledge does not come from experience or that all scientific changes are driven non-empirically. Far from that. The rival models would typically accept that a great deal of justification and revision indeed do come from experience. Where they differ is in claiming that experience is not the only possible and actual source of scientific justification and revision. In this sense non-empirical rival models do not substitute but *modify* the existing dominant epistemic model. Prima facie the intended modification might seem like too big a modification. Yet, from an epistemic point of view to deny the empiricist thesis is not more difficult than to deny any strong claim that pretends to achieve full or almost full generality; and the empiricist thesis is clearly doing so.[4] The requirements that a rival epistemic model should meet are far from easy. Apart from the inhospitable epistemic environment, the model should develop a positive account of non-empirical, that is of a priori justification and a priori revision, and also should show them as actually working in the history of science.

Michael Friedman's model of scientific knowledge represents probably the most influential recent case of addressing the skepticism problem by denying the empiricist thesis. Based on important features of Reichenbach's, Schlick's and Carnap's philosophies of science, the model attempts to combine, after modification, Kuhn's influential view of scientific change with the historically well known neo-Kantian position. Friedman follows the tracks of both Kant and Kuhn and his approach is positioned within a logically-empiricist framework. Since throughout the text I shall *assume* the general model of Michael Friedman I will introduce it, place it in the relevant context and critically analyze it.[5]

The Model of Scientific Knowledge and the Problem of Empirical Revisability

Friedman puts forward a complex three-layered model of a dynamical system for scientific knowledge.[6] The structure of the model is presented by the following three levels:

> *Surface* – concepts and principles of natural sciences: empirical laws of nature, like the Newtonian law of gravitation or Einstein's equations for gravitational field. Faces tribunal of experience by means of empirical testing.
>
> *Second level* – constitutive a priori principles. Defines the fundamental spatiotemporal framework, within which only the formulation of empirical laws and their testing is possible. The principles constitute Kuhnian paradigms – a relatively stable set of rules of the game that allow for problem solving of sciences and the formulation and testing of empirical law candidates. In conditions of conceptual revolution, these are the principles that change under empirical pressure and findings. In periods of revolutions, no empirical testing of them is possible.
>
> *Third level* – philosophical meta-paradigms, meta-frameworks. Guiding, motivating and sustaining the transition between the paradigms (conceptual frameworks).

He argues that the relativized a priori principles accommodate conceptual revolutions and that in fact the revolutions themselves have revealed that our scientific knowledge has foundation layers of such type. The revision of the frameworks requires expansion of our space of intellectual possibilities to such extent, that mere direct appeal to empirical evidence is not relevant during the revolutions. The philosophical and constitutive layer guides the articulation of such new space of possibilities. Therefore, the various levels of our total (scientific) beliefs are not distinguished by mere degree of epistemic security or Quinean degrees of centrality but by their different and still complementary contributions to the total development of scientific knowledge.

	Inhabitants	Properties
Meta level	Philosophical meta-paradigms or meta-frameworks	Serves as a source for suggestions and guidance of the transition from one framework [paradigm] to another.
Second level	Constitutively a priori principles: basic principles of geometry and mechanics. Define the fundamental spatiotemporal framework within which the formulation and empirical testing of base principles is possible Mathematical principles: Euclidean geometry, the geometry of Minkowski space-time, the Riemannian theory of manifolds Particularly fundamental physical principles: Newton's laws of motion The light principle The principle of equivalence.	In periods of deep conceptual revolutions they change "under intense pressure, no doubt, from new empirical findings end anomalies".
Base level	Concepts and principles of empirical natural science; empirical laws of nature like the Newtonian law of gravitation, Maxwell's equations of electromagnetism, Einstein's equations of gravitational field	Face tribunal of experience via rigorous empirical testing

Table 1-1

In this model, each scientific theory (Newtonian mechanics, Special Relativity, General Relativity) has three asymmetrically functioning parts:

a. Mathematical part – contains the basic mathematical theories, representations or structures, intended to describe the spatiotemporal framework in question [*infinite Euclidean space, 4 dimensional Minkowskian space-time, semi-Riemannian manifold*]

b. Mechanical part – in order (c) to succeed using (a) it needs principles of coordination [*Newtonian laws of motion, light principle, equivalence principle*], which set general correspondence between the mathematical part and the concrete empirical phenomena in such a way that empirical laws could have empirical meaning.

c. Physical (empirical) part – attempts to use the mathematical part in order to formulate precisely empirical laws which describe concrete phenomena [*law of universal gravitation, Maxwell's equations for EM field, Einstein's equations for gravitational field*]

Friedman is after two main desiderata. The first one is to preserve commitment to a Kantian or neo-Kantian conception of a priori principles in the exact sciences [logic, mathematics, physics]:

> "… it was not yet clear how one could preserve any kind of commitment to a Kantian or neo-Kantian conception of a priori principles in the exact sciences (as in Kant's original conception of the synthetic a priori, for example, or Rudolf Carnap's version of the analytic a priori developed in the logical empiricist tradition)…" (DOR, p. Xii)[7]

The second desideratum is to account for the dynamics of scientific knowledge. Epistemically that would include fallibility of scientific propositions and the conditions for their revision. The motivation behind the desiderata stems from three main directions: the failure of Kant's original philosophical thesis, the failure of the project of the logical empiricism and the lack of success for the dominant viewpoint:

> "Kant's original philosophical synthesis had failed due to unforeseen revolutionary changes within the sciences, and the logical empiricist's radical revision of this synthesis had also failed to do justice to the very rapid changes taking place within early twentieth century science. " (DOR, xi, preface)

> "… but I was convinced, at the same time, that the dominant view within contemporary scientific philosophy – some or another version of

> naturalistic epistemological holism – is entirely incapable of providing an adequate philosophical perspective on these sciences " (DOR, p. Xii)

Against the Kantian and logical-empiricist background, Friedman argues that we could articulate a conception of dynamical or relativized a priori principles within a historical account of the conceptual evolution of the sciences. He proposes a kind of neo-Kantian-Kuhnian synthesis in order to reach the desiderata:

> "The idea I then came up with, against this twofold background, was that one could attempt to combine aspects of Carnap's philosophy of formal languages or linguistic frameworks with fundamental features of Thomas Kuhn's much less formal theory of scientific revolutions" (DOR, p. Xii).

Friedman develops the notion of relativized a priori principles in various places but recently in the paper *Transcendental Philosophy and a Priori Knowledge: A Neo-Kantian Perspective* and in the book *Dynamics of Reason*. They are supposed to define the fundamental framework only within which the formulation and the empirical testing of empirical laws are possible. A central concern for him is to show that these principles are not empirically revisable. Proof of the opposite would endanger their status as a priori, following a widespread view according to which the a priori epistemic kind is incompatible with empirical revisability.[8] The following passages provide illustration of Friedman's view and the significance of the problem:

> "In constructing his mathematical physics Newton created, virtually simultaneously, three revolutionary advances: a new form of mathematics, the calculus, for dealing with infinite limiting processes and instantaneous rates of change; new conceptions of force and quantity of matter embodied and encapsulated in the three laws of motion; and a new universal law of nature, the law of universal gravitation. Each of these three advances was revolutionary in itself, and all were introduced by Newton in the context of the same scientific problem: that of developing a single mathematical theory of motion capable of giving a unified account of both terrestrial and celestial phenomena. Since all of these advances were thus inspired, in the end, by the same empirical problem … Quine's holistic picture appears so far correct." *(DOR p.35)*

> "Although we explicitly acknowledge that what we are calling here a priori principles (both mathematical and physical) change and develop along with the continual progress of empirical natural science, and in response to empirical findings, we still insist, against Quinean epistemological holism,

> that these principles should nonetheless be seen as constitutively a priori in something very much like the original Kantian sense." *[DOR, p. 71]*
>
> "In periods of deep conceptual revolution, it is precisely these constitutively a priori principles, which are then subject to change – under intense pressure, no doubt, from new empirical findings and especially anomalies." *[Friedman, Michael [2000] "Transcendental Philosophy and A Priori Knowledge: A Neo-Kantian Perspective" in Boghossian and Peacocke (eds.) New Essays on the A priori, OUP; pp. 367 - 383; p. 383.]*
>
> "If these .... Principles ... can thus be empirically tested ... What real point is served by continuing to characterize such principles as a priori?" *[DOR, p. 86]*
>
> "The crucial question, however, is whether such a principle can thereby become empirically false?" *[DOR, p. 87]*
>
> "What can it possibly mean to call principles a priori that change and develop in response to empirical findings?" *[DOR, p. 71]*

The obvious response would point in the direction of the principles being empirically revisable. Friedman needs a contemporary story that settles the problem of the a priority in a modern way. Also, if the constitutivity as a source of a priority is no longer available he would need another positive story about both the epistemic character of the principles and the epistemic nature of their dynamics. In *Dynamics of Reason* Friedman argues that some certain fundamental principles of science are not empirically defeasible because they provide conditions of the possibility of empirical claims and therefore the question of their being empirically false could not arise in the first place. This claim, however, could not stay in isolation. To argue that a proposition is not empirically defeasible is one thing and to argue that it is indefeasible is a completely different one. Friedman does not address directly the question of *epistemic* fallibility of fundamental principles. Yet this question should be addressed since otherwise his argument against the empirical defeasibility could be read in a broader sense to hold against *any* revisability whatsoever and thus could render the principles not merely empirically indefeasible but infallible in general. This would be too much to admit and especially in the face of the current prevailing view in epistemology that all propositions are fallible. Infallibility of the above hypothetically admitted kind would elevate the principles too highly to the status of necessarily true principles; again, this would be too strong to accept and thus it is a clear no-go option.

Relativized A priori Principles and Revisability Prospects

The problem of the epistemic character of the fundamental principles within Friedman's model is crucial for the model as far as it opens it for the empiricist criticism. Apart from the purely epistemic damage such criticism would inflict, it would also undoubtedly change the very structure of the model. For if the fundamental principles turn out to be empirically revisable, they would bear no epistemic distinction with the empirical laws proper anymore, and thus it would remain unclear why and how they could populate an entirely distinct level within the model. The dynamics of reason, which is of central concern for Friedman, would transform into empirical dynamics and the role of the reason would be massively downplayed thus annihilating Friedman's main desiderata: to preserve the Kantian spirit and to preserve a modified version of the Carnapian-Kuhnian structural model of scientific change. Therefore, it is of critical epistemic and structural importance to preserve the a priori character of the fundamental principles on the one hand, and on the other to account for the dynamics, and the dynamics of the principles in particular, in such a way that their a priori nature is kept.

In order to account for fallibility in a modern way, Friedman follows Reichenbach's division between unrevisable and constitutive a priori principles and embraces the view that the constitutive ones are revisable. In this way he avoids the strong objection from infallibility. He argues that unlike the empirical laws of nature they do not face the tribunal of experience. Nevertheless, he concedes, *they change*. What is also significant is that *they do have an empirical content*. In this way we are presented with the following picture: revisable principles that say something about the physical world and which change under pressure from empirical findings, but which are yet supposed to be a priori and not empirically fallible.

Friedman's main line of defense against empirical revisability is to maintain that the a priori principles are constitutive in the sense that they are necessary conditions of the possibility of properly empirical laws. To be constitutive in this sense would mean that if the principle(s) were not available then the empirical laws would not even possess a truth-value and/or would be meaningless. Consequently, the question about their empirical truth or falsity could not arise. In principle, an opponent might attempt to avoid a frontal attack on this constitutive function. However, she could disagree in a different way. Any constitutive principle by Friedman's own qualification does have an empirical component. Therefore, it does say something about the physical world and, in particular, about the way the physical world is. We could suppose that a scientist [or a scientific community] might hold the constitutive principle P, and by doing so she subscribes to some claim about the way the physical world is. There are two options: either she has some reasons to hold that P or she does not. Having reasons is just another way to say that the scientist has some *justification* to hold that P. Clearly, the ”no justification” option is not particularly attractive; most people would prefer to think that scientists are indeed epistemically justified in holding their [coordinative and constitutive including!] principles. In this sense, it is a legitimate scenario where Friedman would have to accept the principles as in a way being justified.

The approach he adopts to secure their a priori character is through the notion of “being constitutive for”; again, it is not clear that this avoids the question about the justification of the principles, and in particular, the question about its epistemic nature. Friedman is explicit about the epistemic nature of the principles; they are a priori, but this epistemic kind should have a clearly identified bearer. The constitutive function of the principles delivers their a priori kind to the bearer: the principles themselves. Yet the relation between the epistemic nature of the a priori and the propositions of the principles is vague; for being a priori is an epistemic property and as such it has to address the question of how it is known or how it is justified. Neither justification nor knowledge is explicitly discussed by Friedman. In the context of the modern epistemic debates it is, however, clear (and especially after the work of Casullo) that the better way of approaching the a priori is through the notion of epistemic justification. For the knowledge requirement might be (and rightfully so) considered to be too strong a requirement, but the fact that knowledge implies justification is more or less uncontroversial. Thus the epistemically minimal analysis that is to be adequate here is much safer to be cashed out in terms of justification than in terms of knowledge. In this way the primary task before the epistemologist is to spell out the relation between the alleged a priori kind and the justification of the principles.

The question about the justification of a given coordinative principle is not merely an internal, in the Carnapian sense, question. For example, when the scientist faces a choice between the old coordinative principles and the new candidates, it does not seem that this could be resolved within the paradigm [or framework]. For Carnap the external questions are not rationally decidable, but in the context of Friedman’s coordinative principles that would translate as the claim that the scientist’s choice is not rationally decidable. Probably a better way to deal with this question is to look for the decidability over boundary coordinative principles [on the edge of two paradigms] within the third layer of Friedman's system of knowledge, namely, the meta-frameworks. Their role is to guide and orient the motivation for transition from one paradigm to another. Again, "motivation" seems to allow that some sort of a good candidate for "justification" is present.

Epistemically, two options might be proposed. The first one is that the justification is pragmatic or conventional and the choice between two coordinative principles is decided conventionally. Historically, conventionality of fundamental scientific principles has been famously defended by Poincare[9] and Reichenbach[10]. It does not seem, however, that in the context of Friedman's model this option is definitive in avoiding the criticism in empirical revisability; since conventional choice made in conditions of intense empirical pressure, findings and anomalies, sounds like just another name for empirical justification. Further, if both candidates to be chosen from are in balanced positions and if an empirical anomaly gives weight to one of them, we could suspect that this in fact might be a kind of empirical revision of the abandoned principle. In this sense, it is not clear how the conventional option would be sufficient for Friedman's purposes. The second option is the empirical justification. This is clearly a no-go option for Friedman, for it endangers the main desideratum, the a priori character of the principles. If a sort of empirical justification regulates the choice between two or more principles this is more or less clear *dependence on experience* and the argued for epistemic kind of a priori would fail to obtain.

Logically, there might be a third option. The third option is that there is a justification and it is a priori. In this case, the situation might not be entirely different. If there is an a priori justification to hold that P, the combination of the principle having an empirical component plus the empirical pressure that participates in the process of causing the change could create difficulties. In particular, it would be difficult to hold that in the exact place, where the principle "touches" through its empirical component the physical world empirical revisability is impossible. The nature of this component, as well as the nature of the requirements we pose with regard to what counts as empirical data or evidence, does not seem to resolve this problem in an obvious way. In principle, whenever we face a claim that says something empirical it is empirically testable if it is so; it might happen that current empirical testing possibilities are limited in some cases but the epistemic nature of the testability remains empirical. As Friedman himself admits, if, for example, the coordinative principle of equivalence turned out to be actually disputed by Eötvös' experiment the principle could not have been simultaneously maintained.[11]

The opponent might interpret this as a possibility that the principle *could have turned out* to be empirically falsified. Historically, this was not the case, but the following hypothetical situation looks perfectly conceivable: let us imagine a coordinative principle, held on exactly the same grounds [whatever they are] as the principle of equivalence and having the same function and success. The only difference would be that the hypothetical principle is disconfirmed by an experiment. Are we still not going to accept that the principle has been empirically falsified? Important consideration here would be that the decision to abandon the principle has been taken solely because of the empirical counterevidence, something consistent with Friedman's counterfactual interpretation of both the principle of equivalence and the light principle. At least hypothetically, we should admit that in such cases empirical revisability is conceivable. Friedman might have a prospective case in affirming that the mathematical part of the theories is not empirically revisable. Nevertheless, he argues that without the coordinative principles the empirical laws of the theory do not have empirical meaning at all. Here, the following question comes naturally: what would deny the opponent the intuitive claim that, even if the separate coordinative principles are not empirically revisable, they are empirically revisable when taken together with the empirical laws? The unity of a priori principles and empirical laws does have a defined empirical content and whether this content complies with the situation in the physical world still seems to be empirically testable. The empirical opponent might also step back and argue for a weaker claim, namely, that given the conditions of empirical pressure and anomalies, this might add to empirical revisability in degree. That is, even if it might not be the case that the leading considerations [whatever they are] for the revision are empirical, there is a certain part for empirical considerations that actually do participate or cause [to a certain degree] the change of the principles. If successful, I believe that even this minimal claim would provide serious reasons for an empirically revisable status of the coordinative principles. In summary, besides the various directions of empiricist type criticism, proper examination of the problem should address the following questions: How much of the a priori would remain if they have an empirical component and respond to empirical findings, anomalies and pressure? How could one back up successfully the a priori character of these principles against empirical revisability? An empiricist might eventually accept their constitutive and coordinative function without at the same time accepting their a priori nature. All of the above taken into account, I will go on to examine what could seem as the most prominent strategy for avoiding the criticism. An a priori defender could put an accent on influential examples in the history of science, and argue that, on the turn of the paradigms, coordinative principles were actually not empirically revised. Prima facie, history of science could provide some good candidates for examples. The core of the strategy consists in the suggestion that even if there is a threat of empirical revisability there exists another kind of revisability which is not empirical and which actually takes place in the history of science. The kind in question is the species of a priori revisability.

The Notion of Constitutive A priori

In 1920 Hans Reichenbach distinguished between two senses of Kant's notion of a priori. The first one is *necessary and unrevisable, fixed for all time* and the other is *constitutive of the concept of the object of scientific knowledge*.[12] The first sense ties intrinsically the epistemic notion of apriority with the modal notion of necessity. The second notion is argued by Reichenbach and recently by Friedman to be the proper notion of a priori, that is, the epistemic notion. The constitutive notion is not examined with respect to its relations to the modal notions of necessity and possibility and the semantic and logical notions of analyticity and syntheticity. Friedman devotes a lot of time to argue that this notion, by the very force of being constitutive of the concept, is empirically unrevisable. However, there are problems with the understanding of this notion that have to be elaborated on. First, all epistemic notions and especially modern ones concern not much more than the nature of our justification and the related questions of *how do we come to know*. If the notion of constitutive a priori is to be such a notion, it has to address this question and should not wander into different domains of philosophical inquiry, such as the modal or the semantic-logical domains. The notion of constitutive a priori however goes far beyond the strictly epistemic. On the one hand it addresses the question of how we come to know by answering: in a non-empirical, that is, in an a priori way. But the notion is obviously a composite one and as such consists in two components. The a priori component is the portion "a priori" that properly answers the epistemic question, but there is another component, namely, the "constitutive" that functions in a different way. The epistemic component is all that is necessary to answer the epistemic question; it is a necessary and sufficient component of an epistemic notion. Every other component that is embedded in the supposedly epistemic notion is redundant from an epistemic point of view and it is not merely unnecessary but also harmful. Every inter-notional component redundancy affects the epistemic address of the proper epistemic component and as such results in a misaddressed component notion in the first place. In this way the first sense of the a priori, according to Reichenbach, is tied intrinsically to the modal notion of necessity. As the subsequent development of the problem of the sovereignty of the epistemic, modal and semantic distinctions has shown, most notably the work of Kripke[13] and more recently of Casullo,[14] all those notions bear a mark of independence and therefore they should not be confused and mixed with one another. The confusion between semantic and epistemic is present not only in the first Reichenbachian sense of a priori but also in the second. It is not difficult to see that, while in the first case Reichenbach (or before him, of course, Kant for that matter), confuses the epistemic notion of a priori with the modal notion of necessity, he confuses the epistemic notion of a priori with the semantic notion of analytic in the second case. The traditional definition of the analytic portion of the semantic distinction that leads its beginning from Kant is that analytic is every statement where the predicate is contained in the subject,[15] whatever this containment notion may turn out to be. Now, the very notion of being constitutive of a concept is an outright notion of containment of the same kind; if the predicate is taken away from the subject the very subject would collapse. In this sense the predicate is constitutive of the subject. Consequently, no proposition containing the subject would be capable of being true or false or meaningful, because the very candidate for a subject in the proposition would be incapable of functioning in it due to the fact that it has been stripped off of the essential predicate. In this sense, all talk about constitutive is actually a talk about analyticity and syntheticity and not a talk about epistemicity. As such, the notion of constitutive a priori is not an epistemic notion but a composite one that features both the epistemic notion of a priori and the semantic notion of analytic. This revealed compositionality bears new consequences for Friedman's defense against empirical revisability.

When claiming that [the concepts of] the differential calculus is constitutive of the concepts that function in the propositions of Newtonian mechanics and also when claiming that [the concepts of] the semi-Riemannian manifold is constitutive of the concepts that function in the propositions of the general relativity, Friedman actually affirms that the propositions which employ the relations between the two are sort of analytic. As analytic the link between the predicate concept and the subject concept cannot be broken without falling into contradiction. This, however, interesting as it is, is an entirely different claim than the claim that the propositions of the differential calculus and a semi-Riemannian manifold are a priori. Because a priori concerns solely how do we come to know and whatever the way we come to know both these say nothing about their revisability. The unaddressed revisability opens the epistemic gap against which the whole edifice of the constitutive a priori has been built in the first place: the empirical revisability gap. Completely aside from the quite plausible claim that neither the concepts of the differential calculus nor those about semi-Riemannian manifolds (let alone them combined with such about physical phenomena) are analytic to the concepts of Newtonian mechanics and general relativity, because we may well deny them without that leading to a *logical* contradiction, the principles in question remain perfectly revisable and nothing seems to make them immune from the danger of empirical revision. Even if it were true that former concepts are so intrinsic to latter that propositions, which attribute a former one as a predicate to the subject of a latter one, would count as analytic [and so, following Friedman, they would have truth value and meaning *only as conjoined*] this does not eliminate the question of how do we come to know the conjoint. Also, it does not eliminate the question about how do we come to know that the conjoint is true [or justified]. This question is essentially an epistemic one and cannot be avoided in any philosophically interesting way through postulating conventionality or similar above-epistemic kinds.

In Need for a New Epistemic Account behind Scientific Principles

Friedman argues about conceptual changes and conceptual revolutions. However, the a priori he needs to defend is not the conceptual a priori, for the empiricist opponent is not attacking it, but the *propositional* a priori delivered through justification. The dynamics he develops as a rational one is a conceptual dynamics, but this does not automatically mean that it is a propositional dynamics as well. The empiricist opponent would question the epistemic nature of the very reasons to hold a scientific principle P or a set of scientific principles, and she could well argue that, even if, say, the conceptual changes were somehow constitutively a priori, this does not mean that the reasons to subscribe to the old or the newly-formed principles are of the same epistemic kind. For subscribing to a concept, she might rightfully argue, is not equal to having reasons to actually hold *a proposition* in which the concept figures. In this way, what Friedman needs is a story about the *a priori dynamics on a propositional level* and not merely on a conceptually-linguistic level. This dynamics is not available in the model and thus it opens the door for empirical criticism in justification.

In what follows, I would like to suggest that such an alternative story is nonetheless available. It is a modern epistemic story, for it directly concerns the epistemic character of the a priori and does not define it via the mediation of non-epistemic notions, be they semantic (like the Reichenbachian notion of constitutive) or modal (like the other Reichenbachian notion of a priori as fixed and unrevisable). It is a propositional story, because it deals with the epistemic nature of the justification behind principles. It is a historically actual story for, as I would argue, we have important illustrations of it in the history of science. At the end, it has good prospects to cope with the epistemic problem of Friedman's model because it is constituted by a strong independence of experience: the fallibility independence. This is the story of a priori revisability which establishes a given proposition or scientific principle, as it is in the present case, as a priori not merely through its being justified a priori, that is, by source apriority, but also through their being a priori fallible, that is, by their revisability apriority. In this story the notion of a priori, as I see it, is much stronger than the logical-empiricist notion of constitutive a priori, as employed by Friedman. For it addresses the modern epistemic debates on the a priori and also, for it solves the problem of fallibility of the a priori fundamental principles in a purely epistemic way with no appeal to semantic or modal argumentation. The a priori revisability of the principles establishes them as (sometimes both) source and revision-wise independent from experience; and in the modern day this is as a priori as it gets. If successful this account of a priori fallibility and revisability should fill in the epistemic gap in the Friedman's model opened due to the employment of the outdated notion of constitutively a priori. Also, besides explaining their actual fallibility it will eliminate the scenario where the fundamental principles turn out to be infallible. In order for the Friedman project to be epistemically successful, the dynamics should indeed be a dynamics of reason and not of experience. The suggested account would add both to the dynamics of the model, through explaining how fundamental principles actually changed in history and to the a priori status of the model, through securing that the change behind the dynamics is indeed, at least in some cases, governed by a priori revisions and not by empirical ones.

The a priori revisability, however, is far from to being a panacea for the rationalist. For it does not disallow other epistemic stories; on the contrary. It certainly does not argue against the existence of or the importance of empirical justification and empirical revisability. It naturally recognizes the role of both empirical justification and empirical revisions and especially so in the sciences. What it does instead is to restrict the scope of the universal empiricist claims as being too strong. It argues that empirical justification and empirical revisability are not *the only* epistemic players and that in a number of cases, some of which fundamental, justification and revision have different epistemic natures. And it does so through an attempt to draw illustrations from the domain that is traditionally considered as the strongest empiricist domain, the domain of science. The nature of the suggested a priori revisability notion is meant to contribute to the moderate rationalist program. It is consistent with a moderate empiricist program, which is liberal enough to abandon the universality of the empiricist claim and to admit that scientific knowledge is the joint product of experience, guided at the fundamental level by a priori principles.

Notes

[1] Kuhn, T. [1962] *The Structure of Scientific Revolutions*, 1st. ed., Chicago: Univ. of Chicago Press., p. 168.

[2] According to the Special Theory of Relativity no signal could travel at a speed greater than the speed of light *c*. The illustration brackets the rich debate on the nature of physical causality and the role of the Light Principle in it.

[3] Tarski, A. [1944] "The semantic conception of truth", *Philosophy and Phenomenlological Research* **4**, 13-47.

[4] One might object that moderate accounts of empiricism manage to avoid the problem as they argue for experience being source of *most* but not all of our knowledge and thus, that it leaves a room for some non-empirical sources. Yet to allow for the conceivability of an alternative source is one thing and to develop a positive account of such a source is a quite different one. Developing the non-empirical sources would be a strange enterprise for any empiricist but also, none of the moderate accounts would attempt to go as far as to develop a non-empirical revisionary dynamics behind knowledge and especially the scientific one.

[5] The current text has been written, revised and prepared for publication *before* Friedman's book *Kant's Construction of Nature: A Reading of the Metaphysical Foundations of Natural Science* to appear and moreover, and unfortunately, before the *Discourse on a New Method: Reinvigorating the Marriage of History and Philosophy of Science* to appear as well. Thus, it fails to account for a new defense of the a priori through the door of revisability, which eventually could be found there. Yet, I believe that the theory of a priori revisability suggested here is still novel and useful and also, I hope that it would help defend Friedman's position against the empiricist criticism.

[6] Friedman, Michael [2000] "Transcendental Philosophy and A Priori Knowledge: A Neo-Kantian Perspective" in Boghossian and Peacocke (eds.) *New Essays on the A priori*, OUP; pp. 367 - 383; p. 383 and also in Friedman, Michael [2001] *Dynamics of Reason: The 1999 Kant lectures*, CSLI Publications, Stanford, p.71.

[7] The quotations in this section are given by Friedman, Michael [2001] *Dynamics of Reason: The 1999 Kant lectures*, CSLI Publications, Stanford (abbreviated in the text as *DOR*).
[8] This view is discussed in more detail in chapters III and IV.
[9] Poincare, H. [1952] *Science and hypothesis*, 1905, New York: Dover. See Especially Chaper III and p. 50.
[10] Reichenbach, Hans [1950] *The Philosophy of Space and Time*, Dover.
[11] Friedman, Michael [2001] *Dynamics of Reason: The 1999 Kant lectures*, CSLI Publications, Stanford, pp. 86 – 91.
[12] The distinction is introduced and discussed by Reichenbach in a separately devoted chapter in his first book from 1920 *Relativitatstheorie und Erkenntnis apriori* (translated in English as *The theory of relativity and a priori knowledge*, Berkeley : University of California Press, 1965). All references are made following the English edition.
[13] Kripke, Saul [1972] "Naming and Necessity", In *Semantics of Natural Language*, edited by D. Davidson and G. Harman. Dordrecht; Boston: Reidel.
[14] Casullo, A. [2003] *A priori Justification*, NY: Oxford University Press.
[15] Kant, Immanuel [1781/87] *Critique of Pure Reason*. Trans. Norman Kemp Smith as *Immanuel Kant's Critique of Pure Reason*. London: Macmillan Co. Ltd., 1963. See *Transcendental Aestetics*.

CHAPTER II

EMPIRICAL REVISABILITY

The chapter prepares the ground for more demanding moves by dwelling on the more or less uncontroversial notion of empirical revisability and stresses its role in science. Analysis of the notion examines and extracts its underlying epistemic structure. The chapter explores the reasoning behind the influential conception of empirical revision of allegedly a priori justified propositions and offers an argument about the possible nature of the relationship between the justificatory epistemic kinds and the revisionary epistemic kinds. The argument presents the empiricist with a dilemma: either to abandon the view that revisionary kinds bear on justificatory ones or to face problems of epistemic identification of her own epistemic kind. Section one presents the notion of empirical scientific revision. Section two discusses the relation between fallibility and epistemic kinds. Section three dwells on whether epistemic kinds bear on each other through fallibility. Section four presents the dilemma of epistemic kinds.

Aristotle famously claimed that the earth is in the center of the cosmos and that the celestial bodies move uniformly with unchanging motion depicting perfect circles around it; Aristotle's description represented a paradigmatic geocentric system. Beyond the nearby celestial bodies, which are the moon, Mercury, Venus, Mars, Jupiter and the Sun are the fixed stars. The subsequent history of science shows that astronomic observations led to changes in this view and eventually led to its substitution. First, the earth appeared not to be in the center of the orbits and the motion of the celestial bodies turned out not to be uniform. Then, the very central position of the earth in the system was disputed. After a complicated history of accumulation of revisionary material and its interpretations that lasted for many centuries, in the early 16th century Copernicus proposed as a substitute a rival system, the heliocentric one, where instead of the earth in the center of the universe there was the Sun.[1] This was a radical change of a theory that describes nature and that has prevailed for a long period of time. Observations and measurement played crucial roles for this and virtually all significant changes within the domain of the natural sciences or so it is accepted by the dominant scientific epistemology, the contemporary empiricism. The hypotheses and theories about the physical world change constantly and this dynamics is one of the core characteristics of all scientific knowledge.

From an epistemic point of view there are two aspects that deserve special attention when it comes to scientific dynamics. The first one is the justification of the hypotheses and theories. The second one is the justification of their revision. Contemporary empiricism tells us that the epistemic nature of both justifications is empirical. While in the case with the former justification, the claim is generally accepted with occasional resistance from rationalists like Friedman and Bonjour, the second claim suffers from an obvious misbalance in the debate: the very epistemic alternative of the empirically justified revision of the scientific hypotheses and theories is hardly if at all discussed. That is to say, virtually no influential scientific epistemologist accepts as a respectable and interesting option the possibility for a non-empirical justification of the revisions in science. The historical roots of this view are probably too long to trace with sufficient precision, but in recent philosophy of science, from the mid-twentieth century onwards, this position undoubtedly carries a significant load from the well known epistemic model put forward by Willard Quine. He notably claimed that *our statements about the external world face the tribunal of sense experience* and also, that *no statement is immune to revision.*[2] The second claim *de facto* legitimates the principle of fallibility and the first one pronounces the empirical revision as the only possible revision; in the natural sciences that is especially meant to be the case. The history of natural sciences seems to provide more than a good confirmation for both of the claims. On the one hand, scientific hypotheses and theories change unstoppably without putting virtually any proposition or set of prepositions on the too-high pedestal of infallibility, and this very fact demonstrates the power of the fallibility thesis. On the other hand, all changes in scientific hypotheses and theories seems to be prompted and justified by observations and measurements of the same kind as the ones mentioned in the beginning of this section. Here I would like to explicate what I take to be the structure of the revision of scientific knowledge as defended by the prevailing empiricism and also, I will point to a special epistemic characteristic of this structure.

The most general revision of a scientific proposition P has roughly the following form:

1. P [affirming something about the subject matter of science S]
2. Q [new proposition(s) due to observation and/or measurement that is related to S]

3. P and Q differ or disagree about the subject matter of S within a scientific framework F [theory + mathematics + logic + available empirical data]
4. P is revised [modified or abandoned] because of or in the light of Q
5. P* or R [that might be adopted within F or might necessitate a new framework F*]

Following this model we might reconstruct in an oversimplified but still formally instructive manner the revision of the geocentric model by Aristotle:

1. P [celestial bodies move uniformly in perfect circles around earth and earth is in the center of the universe]
2. (the set of all) Q(s) [all observational propositions that later would lead to the result R]
3. P and Q differ about the subject matter of the (scientific) theory
4. P is revised
5. R [earth is not in the center of the universe and celestial bodies do not move uniformly in perfect circles around it]

Scientific revisions often take one of the following forms: they either substitute P with R or modify P and reach P*. As it often happens in science R is not all the time easily available or easy to accept; other intermediate steps might be necessary to accept R. Usually those steps involve substantial modification in the initial system F, a modification which, when actually performed, would not allow for F to continue to keep its semantic identity and would result in the identification of a new framework F* that manages to incorporate P* or R in a way better than F does.

The fact that revisions in both philosophy and science do have an epistemic aspect is uncontroversial. In the case of science this aspect, besides being merely of interest for philosophers, has the responsibility to serve as a main tool in establishing the very scientific nature of a proposition candidate for delivering scientific knowledge. Thus, the empirical revisability of propositions is certainly an epistemic revisability; the empirical predicate is an epistemic predicate. If, however, empirical revisions are in fact a subspecies of the more general kind of epistemic revisions, that is, every revision that does have an epistemic aspect, the following question flows naturally: is *every* epistemic revision an empirical one? Contemporary empiricism tells us that yes. The simple fact that there exist an epistemic kind, complementary to the empirical one but epistemically different from it, the kind of a priori, tells us, at least logically, that not necessarily until proved so. It is thus up to the philosophy of science and epistemology to establish whether the logically conceivable non-empirical revisability is also a revisability we encounter *de facto* and such that contributes to the growth of knowledge, be it scientific or not.

Empirically False Vs A priori False

Historically, the modern reaction of resistance towards the notion of a priori has its roots in the dawn of modern scientific principles that were established along with the development of the scientific revolutions, which started with Copernicus,[3] Kepler[4] and Galileo[5] in the 16th century and Newton[6] in the 17th Century. As early as the 16th century Francis Bacon overturned the existing philosophical methods of doing natural philosophy. Until then philosophy was approaching nature by and large through the prism of the deduction, and Bacon's radical proposal was that philosophers should instead use inductive reasoning that starts from facts and reaches axioms and laws. His conception of the new "true and perfect induction" qualifies it as a new and essential tool of the scientific method.[7] Bacon embedded the observation, the measurement and the empirical testing into the new way of doing natural philosophy and their important function became to examine the truth of the general principles or "axioms". The complementary proposal that we should provoke and probe nature where we do not have sufficient information through mere observation is at the heart of the modern scientific conception of the experiment. The subsequent tradition of British Empiricism and the rapid development of modern science followed closely the epistemic tracks of observation, measurement and empirical testing. Even a surface look at the practice and the acting epistemic values of contemporary science would show that current scientific epistemology has not only not deviated substantially from the empiricist framework of justification and defeat, confirmation and refutation, but has even strengthened it. For a proposition or a set of propositions to be scientific it has to be justified from experience either directly or indirectly through inferring from other experiential propositions and, in addition, for a proposition to be *empirically testable* adds quite a bit to its status as scientific, in the first place.

Experience, traditionally linked with truth, is primarily used in modern epistemology as linked with justification, that is, with our reasons to hold that [proposition] P. Yet justification is valuable mostly as far as it relates to truth in order [together with other conditions for knowledge] to provide knowledge. Thus the epistemic expression of the present day is of course "empirically justified" but its more complete and adequate reading should be something like "empirically justified to hold that P" where P, in case of knowledge (and these are of main interest for science) is true and held as a genuine belief by a cognitive agent S, while other conditions for knowledge (whatever they turn out to be) hold too. Obviously, the epistemic load is carried directly by the justification but indirectly, it is carried by knowledge (when there is one) as well. In this way the complementary epistemic expression that mirrors cases of truth becomes automatically visible: where S's justification to hold that P fails to lead to knowledge (be it due to S's actual reasons to hold P, be it due to P's nature or due to, say, the working definition of knowledge) we still keep an epistemic load; in cases where experience is the epistemic type we do not have "empirically justified P (where P is true)" but "empirically justified P (where P is false)". This, however, is not to be mistaken with the mirror expression of "empirically true" – in the whole of epistemic tradition experience was tied with truth – the "empirically false" expression. For easily the expression "empirically justified that P (where P is false)" could be abbreviated to "empirically false" and this would represent a significant inaccuracy in formulation that would result in a serious mistake in meaning: for the epistemic term should only be attributed to what it pertains to and this is the epistemic justification. Thus "empirically false", if to be read correctly, should be read like "empirically *found* to be false" and this is certainly very far from the original meaning of "empirically justified that P (where P is false)". Yet "empirically true" is a pretty much well-formed expression and especially in traditional epistemology, so its counterpart "empirically false" should be well-formed too. The trap is in the ambiguous semantics that rules the expression and allows for two different readings: first, P, which the agent S is justified to hold on empirical grounds and which just happens to be false, and second, P which the agent S is justified *not* to hold and the epistemic grounds of this justification are empirical. It is only the second reading that keeps correctly the link between the epistemic term "empirically" and the justification; the first is simply a misleading abbreviation. For in the first case the epistemic term is not directly linked to the falsity (or the truth, for that matter) of P and thus it is fully independent from the fact that P is actually false (or true). In such cases it is pretty clear that the cognitive agent could not and should not be regarded as having an epistemic route to truth or falsity: she had such route of course, but it was leading in the opposite direction – to falsity where P is actually true and to truth where P is actually false.

This taken into account we can naturally formulate the correct meaning of the mirror expression of "empirically false". Since "empirically false" means "false, and such that we are justified in holding that it is false empirically" its epistemic counterpart "a priori false" should automatically mean "false, and such that we are justified in holding it is false independently from experience, that is, a priori". This keeps the right link between the epistemic term and the justification and opens the formal door for the epistemic formulations of procedures, such as revisions, that are independent from experience.

The standard epistemic objection against a priori justified propositions (or a proposition candidate for delivering knowledge) that runs through the door of their fallibility argues that a proposition P could not have been a priori justified in the first place if it turned out, by virtue of its being empirically revisable, to be false and if the epistemic reasons for proclaiming it false turned out to be empirical. The proposition in question, the objection concludes, is not a priori true but empirically false. Here it is important to distinguish between two major claims the objection is actually making. First, it is the claim of the falsity of the proposition. Second, it is the claim that reidentifies its epistemic status as empirical. Whereas the falsity claim is not surprising as far as it is based on the prevailing doctrine of the principle fallibility of all propositions, the second claim does not rest on a similar generally agreed upon foundation. There is no principle as indisputable as the principle of fallibility that inputs justification (or a proposition that is a candidate for knowledge) of one epistemic kind and given revision due to a different epistemic kind gives as an output a substitution of the initial epistemic kind with the second one. What goes on in the objection, however, is precisely this: the epistemic character of the supposedly a priori justified proposition is being changed or found untenable not because the proposition has been found false but because it has been *found false in an empirical way*. There seems to be an implicit assumption about the relation between fallibility and epistemic kinds. Explicated, this assumption takes roughly the following form:

> If P has an epistemic justification J_a and P is found false through a defeating Q (or set of Qs) with epistemic justification J_b then the strength of the justification J_b defeats the justification J_a and the epistemic kind b that participates in the defeating takes over the defeated epistemic character a.

The leading considerations for such a view apparently accept that the unsuccessful justification of the defeated proposition goes off the board together with its epistemic kind; after all, if the very justification is not successful why should we accept its epistemic kind as successful in the first place?

On Whether Epistemic Kinds Bear on Each Other through Fallibility

An interesting fact reveals the widespread present bias in the current setting in this scenario. Thus, according to the traditional view, the empirical kind of the defeasibility of a priori justified propositions plays a role as a criterion for the epistemic kind of the justification behind the defeasible propositions: if the defeasibility is empirical then the epistemic kind of the supporting justification is not and could not be a priori. Since all propositions are predominantly believed to be fallible it is *the epistemic kind* and not the mere fact of the defeat that purports to re-identify the epistemic kind of the supporting justification. The empirical epistemic kind bears on the a priori epistemic kind but not vice versa. Interestingly enough, however, the a priori epistemic kind has never been dressed up with the same function. The symmetric alternative would be to say that empirically justified propositions are not empirical if found a priori defeasible. Apart from the fact that such a criterion is never proposed the reverse situation is highly instructive about the actual reasoning behind the support of the thesis for empirical bearing on a priori kinds. For it is formally natural *to suppose* that whatever holds for one of the epistemic kinds should hold for the other as well. Let us see how the reverse situation would look like. To take a clear case of empirical justification let us consider the proposition "The cat in front of me has green eyes" (p). In this case I am looking at only one cat in front of me and I see that she has eyes and those eyes are green in color. My entire justification behind p is due to my direct experience of the cat, given all is okay with my perception and my concepts of "cat", "eyes", "green" and "have". Is this propositions defeasible? By all means. How in the epistemic sense p is defeasible, is the crucial question? For if p is defeasible it is epistemically defeasible, that is, the defeater has an epistemic kind and we might inquire about it. Is it empirically defeasible? It surely is. For it might have been a perfect hologram of a cat in front of me and I can learn this by merely putting my hand where the cat should have been; if my hand stops due to the cat's presence there I would have been even more empirically justified than before to hold that p. If not and the hologram ray's projection appears on my hand then I would immediately grasp the real empirical situation and would revise my belief that the cat in front of me has green eyes; for there is no cat let alone one with green eyes in front of me but a hologram and thus p would be false. In this case I have found in an empirical way that p is false and I have empirically defeated p. Is this the only possible epistemic way to question p and eventually, defeat and revise it? Let us see. The only other epistemic way to question p that is not empirical is trivially the non-empirical, that is, the a priori defeasibility and revision.

The apparent trouble with this sort of revision in the case of our p is that it is not at all obvious how we could even approach p non-empirically. What could possibly be the reason for that? Well, for starters, the very proposition comes with a sort of semantics for its evaluation. Standardly, p is true if and only if there is the physical object "me" gifted with the ability to visually percept, a [physical (animate, as it were) object] "cat" having the property of being in front of me and the property of having eyes, the physical property "being green" and all those physically have the properties they are said to have and stand in the way they are said to be by the proposition p. If this were not the case p would not have been true. The semantic setting of this proposition that is dictated by the very nature of the terms and relations employed in p produces the kind of conditions that would render p true or false. Yet those conditions in order to be established for the purposes of the epistemology should be somehow accessible in order to be known whether they are satisfied or not. In other words, the cognitive agent [me] should have some kind of access to the state of affairs that would render p true or false. The state of affairs that would render p true or false is the above described physical state of affairs and therefore I need some kind of access to the physical reality. The epistemic defeasibility that could confirm or disconfirm p should have as many epistemic kinds as the number of available accesses types to the state of affairs that render p true or false. The standard way to access physical reality is through sense-experience and in this particular case, through the sense of vision. Is this the only possible access to the state of affairs that deals with p's truth or falsity? By no means. For example, I could be told that the cat in front of me has green eyes, thus having a testimonial justification for p. Or, I could read a fortuneteller's book which predicts with certainty that on such and such date at such and such place if I stay in such and such position there would be a cat with green eyes in front of me. But all these other types of access pertain to the general kind of empirical access and therefore, varying as they are, they do not constitute an epistemically different kind of access. The hallmark for the empirical type of access is the route of delivery: it could be through sense-experience or enhanced sense-experience, like the one that results via usage of devices like glasses, telescopes, microscopes and the like, or it should be the route of delivery that is ultimately reducible to one of those kinds. Is there any way that we could know if p is true or false that does not rest on empirical kind of access of the above kind? Is there any way that I could trustfully check whether there is a cat in front of me and whether its eyes are green without appealing to any sort of empirical access to the state of affairs that would render p true? It does not seem so. I cannot check purely logically whether the thing is a cat or not and whether its eyes are green or not; in order for logic to work in this case it needs to operate on information that is already present and the obvious way of delivering this information is through experience. I also cannot purely mathematically approach the truth conditions of p, nor I could infer somehow modally that *there should be* a cat in front of me with green eyes for it could not have been otherwise; it sure can. Yet these are traditional domains that purport to deal with non-empirical justifications and eventually, to deliver a priori knowledge; all of them seem helplessly irrelevant to evaluate p. For even the rudimentary, so to speak, check whether the thing before me is a cat and not a cleverly disguised dog which identifies the very object referred to by the term "cat" is impossible if there is no some sort of causal connection with the object and in the general case this causal connection could be traced down by means of some [sense] experience. It is not a deficiency of the non-empirical access, whatever it might turn to be, that it is not able to reach for the state of affairs that would render p true. It is *the very nature* of the truth conditions of p that determine which access is metaphysically possible. And clearly, in the case with the cat it is the empirical access that is the only possible one; any non-empirical access does not seem conceivable at all.

Thus, it seems that for some propositions like the above p, which employs terms whose referents seem solely empirically accessible, there is only one epistemic kind of defeasibility and this is the empirical defeasibility. If this is correct, how does the question about the epistemic fallibility of p strike us in the first place? How does the possibility of p being found false in an a priori way sound? And most importantly, how does the possibility for p to turn out to have been a priori justified because it has been found false in a non-empirical way? For it should be clear now that p could not have been revised a priori, that is, p is *not* a priori defeasible let alone a priori revisable and thus ending as non-empirically justified due to the rival epistemic kind of the defeating justification. Certainly, this does not mean that *all propositions* justified by one epistemic kind are not revisable by propositions of the other epistemic kind; far from it. But it demonstrates that *in some cases* the very epistemic nature of a proposition makes it inaccessible from the position of the alternative epistemic kind. It is pretty clear that some empirical propositions are not opened for a priori revisions and in those cases such revisions are practically inconceivable. The referents of the singular terms of the cat proposition and in fact the vast majority of perceptually justified propositions are simply such, that non-perceptual assessment for the success of their reference is inadequate.

The natural symmetrically translated step is now easily visible: if some empirical propositions are a priori unrevisable what about some a priori propositions being empirically unrevisable, too? Clearly, there are propositions like those of mathematics, whose subject matter is radically different from standard perceptual and observational propositions. Thus it does not seem unnatural to suppose that the epistemic access for both their justification and revision is *different* from the empirical one. Propositions like the above empirically delivered p all seem to follow in those tracks: if a proposition is rendered true by a state of affairs that is accessible only and solely by experience then these propositions are only empirically fallible and not a priori fallible. In these cases the epistemic kind of the defeating justification does not and it is not able to bear on the epistemic kind of the supporting justification at all.

The problem for the prevailing view now becomes evident: why not apply the same type of reasoning in the case with the reverse situation where an a priori justified proposition is being empirically revised? What if the state of affairs that would render an a priori justified p true or that would contribute to the strength of its justification is not accessible through any experience whatsoever? Should this lead us to deny that the propositions in question have semantics in the first place? To illustrate, would this justify us to accept that propositions about logical and mathematical truths are semantically void or isolated? For this does not seem to be the case with our logical, mathematical, modal and scientific practice. Logical truths seem unavoidably true and it is this inevitability that makes them so precious and useful in the first place. For if there was no obvious appeal to believe that if *a* is bigger than *b* and *b* is bigger than *c*, then *a* is bigger than *c* and especially, to believe that *this actually is the case*, then logical truths like this would have been long abandoned in our practice. This, naturally, is very far from the rational human practice. If the proposition $2 + 2 = 4$ was not describing actual truths that do hold about all possible referents under its scope, numbers or numbered objects, it would have been of no use whatsoever for both mathematics and everyday practice and the mathematical system that employs it would probably not have been formed in the first place. Or, to continue, we should conclude that because the semantics which those propositions have is not easily accessible like the ordinary semantics of empirical propositions, then it is a bad and ill defined semantics which is problematic. Even if this were the case it would not follow that we are not justified a priori to hold some propositions and if we are so justified, fallibility granted, it is perfectly legitimate to approach to defeat them.

A dilemma seems to emerge: either epistemic kinds do not bear on each other through defeasible justifications or there is an asymmetry between the epistemic kinds and whatever rules hold for the a posteriori do not hold for the a priori and vice versa. Apart from the obvious topological kind of considerations like difference in complexity and simplicity, bigger versus lesser aesthetic appeal and the like, this dilemma does not seem easily decidable. It is true that prima facie the space of epistemic possibilities should comply with some general structural rules and symmetry seems like quite good a candidate for one of those rules. It is also true that it is not clear why we should accept different rules for one of the kinds that put it in a sort of primacy position with regards to the other kind and especially, it is not clear why we should do so due to considerations that are backed up almost entirely via the very epistemic kind to be found dominant; for all the reaction against epistemic symmetry is due to the faith in the delivery route of the empirical kind and yet, again, this route is as insecure as it gets. This, however, sketches only one of the sides of the problem. On the other side stay questions like what if the very nature of the subject matter of a priori and empirical propositions turns out to be of a different kind; then it would not seem unreasonable that different rules of the game determine the space of epistemic possibilities and thus the empirical kind could bear through defeasibility on the a priori but not vice verse. Or, what if correspondent semantics, turned out to be different, determines the rules and so symmetry is broken again?

In approaching the dilemma we should distinguish between function and content. The content of a proposition is accessible in some way or another and in this sense it is clear how the two epistemic kinds differ: they simply pertain to propositions with radically different epistemic content types. This however is very different from the function which an epistemic kind performs. For delivering the information is one function but serving as a criterion for another epistemic kind is a very different function. No question about the possible relationships between types of justification and fallibilities might be even approached if the epistemic kinds are functionally not well defined. The current debate over the empirical defeasibility of a priori propositions is a good illustration for this. And their functioning as a criterion is a definite part of this correct definition. Let us see how the functional space of the epistemic kinds would look like if try to define it asymmetrically with the empirical kind serving as a criterion for the a priori one and not vice versa. Purely formally, an epistemic kind is constituted by the nature of the route that delivers some sort of information. With respect to this both kinds are on a symmetrical par. Further, in order for the distinction between the available kinds to be non-redundant and meaningful the kinds *should differ sufficiently* so that one does not be taken for the other. Third, the space of validity of both kinds should be sufficiently well defined so that no candidate for an inhabitant of the space drawn from one type of the candidates could be found in the space drawn by the other type. These requirements constitute what seems to be a necessary part of the definition of multiple [at least two] epistemic kinds. Yet all these formal requirements carry the mark of the symmetry and all kinds are treated under the same restrictions.

The question is: is it a legitimate part of the definition of an epistemic kind to be able to serve as a criterion for identification of another epistemic kind? The main criterion in judging that could only be the correct definition of the kinds involved. And here the asymmetry shows how a function of this kind leads to formal problems with the whole epistemic functional space. For if a kind *a* is a criterion for a kind *b* within the space, then they are hierarchically ordered and not horizontally. That would mean that both kinds do not treat their subject matter in a symmetrical way. Yet the subject matter of the kinds is nothing above the route that delivers information. How could two routes differ in a hierarchical way? How could route 1, which is a different route from route 2, serve as a criterion for what route 2 actually is? And most importantly, how does route 1 get its proper definition in the first place? Does it get it from route 2? This option is not available, to take it for the sake of the argument. Then, route 1 should either rely on itself to define itself, which, if allowed, would change it from a route to a sort of a rule or, it should rely on some previously and therefore independently defined rules that would qualify it as a route and not as a rule. The obvious problem here would be a difference in kind: if route 1 is allowed to define itself then it would differ from route 2 not only with respect to its function but also with respect to its kind. For route 2 would not be a rule but a route. And yet this is like nothing we observe in practice: epistemic kinds are both routes of information and thus they do not differ in kind, they are both epistemic kinds, but only in content which is necessary to distinguish between the two. Therefore, the possibility for route 1 to be able to serve as a criterion for or self-identify itself does not seem a good possibility and looks much more like a circular move. The only option left then is route 1 to receive its identification through independently defined rules that first attribute which route has which function. Here again we see nothing but symmetry in the requirements that provide sufficient differences between the one and the other epistemic kinds. And if symmetry is part of the rules, then, all routes should be symmetrical with respect to their functions. If this is correct any rule should have every other function as every other rule and if it does not, they should not be contrasted as being of the same type. If this is the case then no kind could serve as a criterion for the identification of any other kind and this means that neither the empirical nor the a priori kind could bear on the other kind whether or not through defeasibility or otherwise the kinds would end up as being ill defined.

A final point throws additional light on the epistemic mechanism that regulates the relations between the epistemic kinds of the justification of an empirically justified p and the epistemic kinds of the revision performed. If we take our p, empirically justified as it is, and ask what would happen if we revise it empirically all empiricists would happily agree that at the end of the revision we would still have an empirically justified proposition. Their, implicit perhaps, reasoning is that because of the empirical epistemic kind of the revision, the epistemic kind of the proposition is affirmed as empirical. However, the true epistemic kind they actually refer to is the epistemic kind of the justification behind the proposition. Thus, it is clear that what in fact the empiricist tacitly accepts is that the epistemic kind of the justification is determined post-factum, so to speak, by the epistemic kind of the revision of the proposition. This, however, is far from obvious. For, and this is the interesting moment, no empiricist would ever admit that an empirically justified proposition would turn out to be a priori (justified) *were there cases* where empirical propositions are indeed a priori revised. Thus, it seems that the empiricist is using a double standard: the first one says that empirical epistemic kinds are preserved no matter the epistemic kinds of the revision and the second one says that the a priori epistemic kinds are lost if the epistemic kinds of the revision are empirical. The obvious problem with this is that it looks terribly *ad hoc*. The mechanism that regulates the relations between the epistemic kinds is an epistemic mechanism and as such all epistemic kinds in play should follow the general rules of epistemology.

The Dilemma of Bearing Epistemic Kinds

If the empiricist chooses to have as a valid scenario the one where epistemic kinds actually do bear on each other she could retain the force behind the standard empiricist argument against the a priori. However, she might potentially lose something too for now the epistemic kind of the a priori would have the ability to bear on the epistemic kind of the a posteriori, as well. On the other hand, if the empiricist chooses to have as a valid scenario the one where epistemic kinds do not bear on each other, she would lose the force behind the standard empiricist argument. Thus the empiricist faces a dilemma where she should chose between two arguments. She faces either the

Argument for trans-epistemic penetrability

1. according to a prevailing view a priori propositions are empirically revisable
2. empirical (E) and a priori (A) are different epistemic kinds
3. if (1), then E does have some bearing on A by virtue of fallibility of proposition A and by virtue of the actual revision
4. If (3), epistemic kinds are capable of commuting with each other and also
5. one epistemic kind could epistemically annihilate another one
6. if (5) *and* epistemic symmetry, then a priori annihilation of empirical propositions is conceivable

Or the

Argument against the epistemic loss of kind due to revision of opposite epistemic kind

1. suppose that an empirically justified proposition is revised a priori
2. would this change its justification from empirical to a priori? No.
3. if not then the epistemic kind of the revision does not bear on the epistemic kind of the justification
4. both epistemic kinds are revisionary independent from each other
5. if (4) and (5) and epistemic symmetry, then
6. empirical revision does not bear on a priori justified propositions as well
7. fallibility is independent from epistemicity

I do not believe that the choice for the empiricist out of this dilemma would be a very difficult one. Being able to attack forcefully your opponent is certainly important but not as important as the danger of exposing one's own position to the vulnerability of being potentially deprived from its own epistemic kind. It is very unlikely that an empiricist would concede that it is possible for an empirically justified proposition to lose its epistemic kind just because it is a priori revisable or worse, a priori revised. Therefore, having epistemic symmetry as a guide, we have to acknowledge first, that the empirical sorts of justification, fallibility and revisability are giving just half of the epistemic picture. Thus a new, more abstract conception of fallibility and revisability should be considered and these are the general kinds of epistemic fallibility and the epistemic revisability. In this new context empirical fallibility and empirical revisability need their subsidiary epistemic counterparts of a priori fallibility and a priori revisability. The conceivability of epistemic revisability is, however, not sufficient for the apriorist. From an epistemic point of view she needs to show actual cases of a priori revisions in order to establish it as a genuine epistemic mechanism. From the point of view of philosophy of science she needs to demonstrate that a priori revisions indeed do happen in science. For our present purposes both desiderata are needed in order to defend epistemically Friedman's model of scientific knowledge. Thus the next two chapters will dwell on the notion of epistemic revisability and the newly conceived of notion of a priori revisability whereas the last two chapters will propose two historically influential cases of revision in mathematics and physics as an illustration of the actuality of a priori revisions in science.

Notes

[1] Copernicus, Nicholas [1543] *De revolutionibus orbium coelestium* (On the Revolutions), Nuremberg, 1543 C.E.

[2] Quine, W.V.O. [1951] "Two Dogmas of empiricism" in *The Philosophical Review* 60 (1951): 20-43.

[3] Copernicus, Nicholas [1543] *De revolutionibus orbium coelestium* (On the Revolutions), Nuremberg, 1543 C.E.

[4] Keppler, Johannes [1618 - 1621] *Epitome astronomiae Copernicanae* (Epitome of Copernican Astronomy) (published in three parts from 1618–1621)

[5] Galilei, Galileo [1632] *Dialogue Concerning the Two Chief World Systems* (1632, in Italian Dialogo dei due massimi sistemi del mondo) and *Two New Sciences* [1638] (in Italian, Discorsi e Dimostrazioni Matematiche, intorno a due nuove scienze)

[6] Newton, Sir Isaack [1687] *Philosophiæ Naturalis Principia Mathematica* (from Latin: "Mathematical principles of natural philosophy"),

[7] Verulam, Lord Francis [1898] *Novum Organum* or *True Suggestions for the Interpretation of Nature*, London and New York. Book II.

CHAPTER III

EPISTEMIC REVISABILITY IN NATURAL SCIENCE: EMPIRICAL VS. A PRIORI

The chapter develops a notion of epistemic revisability and argues that it should be regarded as coming in two kinds, empirical and a priori, even in natural science. Section one introduces the prevailing view in current epistemology that all or most statements are fallible (the principle of fallibility) and explores the epistemic nature of the relation between justification and revision. Section two formulates the Empirical Revisability principle. Section three introduces a distinction between revision of truth-value and revision in meaning. Section four recalls the problem of the bearing of revisionary epistemic kinds over justificatory epistemic kinds. Section five argues that the prevailing conception of empirical revisability leads to an undesirable misbalance between the epistemic kinds. Section six stresses the importance of the symmetry between the epistemic kinds. Section seven discusses the difference between the notion of conceptual a priori and the notion of justificatory a priori.

Fallibility, Empiricism and Revisability

Among the widely accepted views in contemporary epistemology is the one that all non-elementary logical and mathematical truths are fallible, or that in principle, every statement could turn out not to be true. Fallibility has been an influential stance throughout 20th century philosophy; also, the view has been especially forceful in the domain of the natural sciences. A notion complementary to and inseparable from the notion of fallibility is the notion of revisability. If a proposition is fallible then it is revisable. There should exist some sort of epistemic procedure via which the strength of its justification or its truth or falsity is examined. In the general negative case, when the proposition is disconfirmed and eventually substituted with a different one we accept that the proposition has been revised. Such negative cases present the formal type of revision of proposition P without the epistemic aspect of the revision being taken into account. A symmetrical analogy with the epistemic aspect of justification would provide the simplest yet the most natural definition of the epistemic aspect of the procedure of revisability.

The epistemic aspect of the nature of justification is standardly taken to be provided by the answer of the question "What is the *epistemic nature of the reasons* to hold that P?". Symmetrically, the epistemic aspect of the nature of revisability should be provided by the answer of the question "What is the *epistemic nature of the reasons* to revise P?". Traditional epistemic choices to pick from are experience and lack of experience. The former provides the current widespread epistemic basis of the experiential nature of all or most of our knowledge, the *a posteriori* one; the latter provides the significantly less popular epistemic basis of the non-experiential nature of some of our knowledge, the *a priori* one. The domain of natural sciences is a traditional fortress of empiricism whereas the domains of logic and mathematics are typically taken to allow for non-experiential knowledge.

The standard way to define epistemic justification is to say that S is justified to hold that P if and only if S has some reasons to hold that P. The segment *having reasons to hold that P* conveys the core idea of the concept of justification. Once the concept is accepted in this sort of minimal way a minimal concept of revisability flows naturally. The complementarity of justification and revisability leads to further symmetry of definitions

a. justification $=_{def}$ having reasons to hold that P
b. revisability $=_{def}$ possibility of having reasons to abandon P

Whenever there are *reasons to hold that P*, fallibility taken into account or not, there could always be *reasons not to hold that P*. The reasons in the case of justification are usually taken to be direct and to contribute directly to maintaining that P. The reasons in the case of revisability seems to be of a different sort. Having reasons not to hold that P does not necessarily contribute in a direct way to the eventual revision performed on P. The revision might or might not be performed and yet the fact of the presence of the reasons not to hold that P seems to be independent of the actual revision. For the present purposes I would restrict the analysis only to the species of epistemic revisability within the natural sciences and what is often taken as a core fundament of theirs, mathematics.

The influential contemporary view on epistemic revisability in natural sciences is with no doubt the empiricist one: revisions in sciences are prompted, conducted and justified through appeal to experience. Thus, *de facto* in the domain there exists just one epistemic type of revisability, empirical revisability. Current epistemology, although well engaged with accompanying issues that come with the idea of empirical revision is as a rule strangely silent about possible epistemic alternatives. An important condition that has to be kept in mind, however, when it comes to "non-empirical evidence", is the following: although the evidence is explicitly said to be non-empirical it does not automatically follow that it is epistemically well defined, let alone being of the rival epistemic category, namely, the a priori. For in order for evidence to be considered as a priori and not merely non-empirical (and besides the apparent lack of informativeness behind such minimal negative qualification), there has to be an epistemically coherent conception about what could count as an a priori in the first place; and second, the suggested non-empirical evidence has to be conclusively shown to be of that kind both in an epistemological way and via an appeal to actual cases. Even a fast look at the current debate on the a priori demonstrates that the first requirement proves to be a highly problematic one. Apart from the obvious high tension of disagreement between the available positions which makes it practically impossible to follow a generally agreed upon informative notion of a priori, for the lack of such notion, there is another problem which is even more severe. The problem is that there could be no satisfactory coordination of a priori candidates with actual cases if no account on the a priori is available and if the acting notion is superficial or colloquial (as it often is the case) and not an epistemically legitimate one. Surely, the traditional notion of a priori as knowledge or justification that is independent from experience is easily available but this notion is not very helpful until the nature of the independence is spelled out in sufficient detail. Also, this prevents such non-empirical accounts of evidence from a concrete full blooded defense of having actual reference to historical events and especially so in the sciences.

There are quite a few suggestions for "non-empirical evidence" in the relevant literature and yet there are no epistemically viable accounts that are presented in an affirmative manner. Standard candidates are logical deduction, logical and mathematical intuition, thought experimenting. To illustrate, in his discussion of the fallibility of a priori knowledge Aron Edidin explicitly points to logical deduction, logical intuition and thought experiments as paradigmatic cases of non-empirical evidence.[1] Yet in order to be able to use thought experimenting as illustration here he would need an account of *why* thought experiments are a priori. Such an account, however, apart from being rarely offered in the literature on thought experiments, (James Robert Brown's recent platonic view here would be a notable exception)[2] and apart from being under immense pressure from harsh empiricist criticism, (John Norton's influential position being the paradigmatic example here)[3] could not be simply presupposed. In order to become available it needs a notion of the a priori backed up with illustrations from history of science. Even if we accept Brown's a priori notion as *prima facie* epistemically sufficient, the problem of the fallibility of a priori knowledge and justification remains unsatisfactory resolved and the empiricist criticism should be met with respect to a reformulated notion of a priori that accounts successfully for the fallibility of the a priori propositions. After Edidin, Albert Casullo and Joshua Thurow, among others, have recently entertained the idea of non-experiential evidence. Thurow argues that many a priori beliefs are defeasible by non-experiential evidence[4] and he draws support through illustrations of fallibility of mathematical justification; yet his proposal is vulnerable to an epistemic objection which again concerns the nature of the relation between the notions of a priori and fallibility. Thurow's main thesis is that those a priori propositions which are defeasible (because he accepts the existence of a priori propositions that are indefeasible) even if defeasible by non-empirical evidence are also defeasible empirically. The problem is that empirical defeasibility is in no way helpful for identifying a proposition as a priori. Philip Kitcher points here to a danger by arguing that if a priori propositions are empirically defeasible then they would be practically indistinguishable from the empirical propositions.[5] Even if we do not have to accept that empirical fallibility would completely annihilate the a priori it is obvious that it would deflate the notion of the a priori to such a degree that it would not serve any significant philosophical interest whatsoever. Therefore, if we agree with Thurow's implication that if an a priori proposition is non-empirically defeasible then it is also empirically defeasible, we would end up with useless and thus uninteresting notion of the a priori. Yet a huge portion of the epistemological interest in domains like mathematics and logic stems from the clear epistemic difference in kind between those domains and the domain of the natural sciences. Thurow's own admittance of indefeasible a priori propositions would not help much here too; for if they are indefeasible they are also empirically indefeasible and thus their a priori status, given the regulatory function of the empirical defeasibility, would turn out to be different then the a priori status of the empirically defeasible a priori propositions. The apparent apriority of a mathematical proposition like $2 + 3 = 3 + 2$ and a logical truth like "if a is bigger than b and b is bigger than c then a is bigger than c" does not seem to be of a different kind, does not seem to vary in the first place and in particular, it does not seem to vary due to empirical fallibility. For the reasons to hold both types of propositions do not rest in an obvious way on experience; no appeal to experience seems to be capable of delivering the truth the propositions allegedly have and consequently, no appeal to experience seem to be able to *diminish* the justificatory support. The very relation between experience and mathematical and logical truths is notoriously difficult to come up with let alone to exploit in order to back up supporting or defeating justification. Thus, to deflate the notion of a priori would cause an epistemic vacuum to account for the difference in a priori within mathematics and logic; in the face of the alleged apriority of mathematics and logic such a vacuum does not look promising to fill in.

The Principle of Empirical Revisability

The fallibility of propositions thesis that prevails in contemporary general epistemology is especially dominant in the more specific domain of scientific epistemology. It follows directly from the fallibility thesis that a revision of a proposition, either in meaning or in truth value, is practically always conceivable. It is also implicitly understood that the result of the revision, as far as it is based on a strong argument, is meaningful and eventually, better justified than the revised one. The backbone epistemology of contemporary science is modern empiricism. Together with the fallibility thesis this means that whatever reasons there are for a statement to be justified it may come only through experience. By exclusion, the same should hold for any reasons to revise a scientific P as well: no empiricist would accept both that on the one hand the only possible source for justification is experience and on the other hand that for the revision of a justified proposition there is also an additional, alternative source, like an a priori one. Every revision of a proposition represents some form of an argument that contains among its premises the proposition, subject to revision and has within its conclusion the revised (or fully substituted) version of this proposition. In this sense, the very act of revision of P is itself nothing but a justification of a different statement, which is either incompatible with the revised one [in cases of radical revision] or it just represents a shift in its meaning. As far as it is justified it has a possible domain of reasons that would back up its acceptance. Thus, the epistemic question of the possible source of reasons that back the revision translates naturally to and reduces itself to a standard question about the epistemic nature of the reasons to hold that P, that is, the epistemic nature of the justification for P. Together with the empiricism thesis this renders the following picture [Empirical Revisability Principle, ERP] :

(ERP) Every revision is a justification [of a different statement] and as such [because of empiricism] it could only have an empirical nature for the reasons that back it up.

Revisability in Truth-Value and Revisability in Meaning

In their seminal paper "In Defense of a Dogma" Grice and Strawson distinguish between "shift in sense of the words" and "change in truth value".[6] The distinction is particularly instructive in the context of revisability for it points to two possible ways of revising a proposition. The first kind of revisability is the standard one that is presupposed in most discussions about revisability and concerns the revision of the truth value of a proposition. In this scenario, taken from within scientific epistemology, a proposition P taken within a theory and having a truth value T_1 undergoes revision and ends up with another truth value T_2. It is important to note that this type of revision does not tell anything about possible functioning of the revised proposition outside its theoretical system or within a different system. The only aspect that it addresses is the change of the truth value. Certainly the change of the truth value bears serious consequences for the proposition as an element of a theoretical system. This revisability, however, does not exhaust completely the types of the available revisions and especially the revisions in science. Very often given proposition P although suffering from a revision does not undergo complete substitution with a different proposition Q but it is still modified in some way. There is another aspect of scientific revisions that is left unaddressed by the standard truth-value revision. The *meaning* of the revised proposition undoubtedly changes after a revision whether or not the proposition changes its truth value. Therefore, a different type of revisability should be defined that addresses this change in meaning. Thus, the second kind of revisability is revisability in meaning. In this scenario a proposition P taken within a theory and having a meaning M_1 undergoes a revision and as a consequence changes its meaning from M_1 to M_2; as a result, the proposition P transforms into the proposition P* which has a shifted meaning. We have to distinguish between two possible sub-scenarios. During the standard meaning revision P more or less manages to keep its identity because the shift of the meaning is still compatible with the initial proposition and in a way just modifies it. In a severe case the shift of the meaning is such that it is incompatible with the initial proposition's meaning and leads to a different proposition that would be the bearer of the new meaning. In this case we do not have a transformation of P to P* but rather, we have a substitution of P with Q. It is natural to suppose that most radical changes in meaning would simultaneously lead to revision of P's truth value as well, that is, by virtue of the great shift in its meaning P would lose its initial truth-value T_1 and it would receive a truth-value T_2. But this might not necessarily be the case.

The distinction between revision in truth value and revision of meaning might not seem entirely well defined due to the ongoing debate about the relation between truth value and meaning; however, the basic notions of truth value and meaning of a scientific proposition seem nevertheless sufficiently well defined so that they could be distinguished in the first place. Generally, in epistemology (but especially in science) whether a proposition is true or false or better or not so well justified is quite different an issue from the issue of the same proposition being meaningful or meaningless. Certainly the nature of the relationship between truth value and meaning are of utmost philosophical interest but for the present purposes, apart from the fact that it is practically impossible to shed sufficient light, emitted from a prevailing consensus, such an investigation is not necessary in order to make use of the distinction. A proposition like "Earth is in the center of the Solar system" is, as we know today, false, according to current science but it is still a perfectly meaningful proposition. On the other hand if a scientific proposition is to be true (like "[According to our best theories] The Sun is in the center of the Solar system") then it is inevitably meaningful. In this sense, in order for a proposition to start to be capable of having truth values as "true" or "false" it has to be meaningful in the first place and the reverse relation apparently does not hold; a proposition does not need to be settled as true or false in order to be meaningful. Such remarks which just touch upon the surface of the problem are instructive enough about the actual distinction between truth value and meaning which is found in today's scientific practice. For a proposition to be well defined scientifically it has to comply with the current scientific standards for proposition formulation; that would include usage of the scientific observational and/or theoretical terms, compliance with the current theory hosting the problems the proposition address, meeting certain standards of rigor like compliance with certain logic, that serves the scientific theory and compliance with some mathematics that formulates the propositions of the theory. Failure of the proposition to meet such requirements would affect its meaningfulness. Meeting the requirements would grant the proposition its status of a well-formed proposition within or with respect to a certain scientific theory.

A related problem is the relation between a truth-value of a scientific proposition and the justification of the proposition. It is generally accepted that justification does not entail truth.[7] In this sense it is epistemically much safer to speak about justified propositions and not about true propositions. In what sense then the distinction between revision in truth-value and revision in meaning could turn to be useful? If all the major epistemic considerations are cashed out as a talk about proposition's justification then the issue of its truth-value seems to be circumvented. This question looks quite legitimate but it faces different epistemic situations that correspond to the scenarios of general epistemology and scientific epistemology. In the general epistemic case a proposition might perfectly well be epistemically discussed on its own and not necessarily within some system of propositions. We can always ask about the truth or the meaning of the proposition "There is a tiger on the mat" and in order to establish this all we have to do is just look at the mat armed with our concepts of "tiger", "mat" and "is". In the general case we do not seem to need a set of additional propositions only in the light of which the above proposition could turn out to be true.

In the case of science, however, the epistemic situation is different and usually in the scientific practice no proposition is regarded as true or false in isolation but always within a system of propositions or a theory. Scientific theories develop constantly and this results in a permanent change of some of the propositions in the theory. In a restricted sense the scientific community regards all scientific propositions as having their truth value relativized to a certain scientific theory. Thus, the truth value could not be regarded as an absolute one, i.e. as a truth value that would hold, for example, for another theory or in a different epistemic situation as everyday talk. A proposition formed within Newtonian mechanics does not necessarily have the same truth value within, say the General Theory of Relativity, for this proposition could have been regarded as "true" in the former but would most probably regarded by the scientific community as "false" in the latter theory. What is also a related fact that deserves attention is that a proposition with a truth value taken within a theory does not necessarily have to be meaningful outside the theory in the first place. As Friedman argues, Einstein's equations of gravitation from the general theory of relativity, true as they are regarded today within the theory, are not regarded as even capable of being false within a different theory like Newtonian mechanics; because they could not possibly have been formulated in a meaningful way.[8] The mathematical theory that allows for their formulation, the Riemannian theory of manifolds, is not part of the Newtonian mechanics and an attempt at its incorporation would radically change the claims of the theory. Also, the fundamental principles that set an important framework for the theory, like the principle of equivalence or the light principle, do not figure in the old theory and this is crucial for the equations to receive meaning in the first place. In this sense both truth-value and meaning are intimately connected to a certain scientific body of theory or theories, and for a scientific epistemology it is important to account for this fact. Does this mean, however, that there is no epistemically interesting truth-value of scientific propositions that would play a role in the above mentioned revisions in truth-value? Not really.

There seem to be three basic types of truth values that a proposition could have when taken within a scientific theory. First, the proposition could be regarded as true within a theory. Second, it could be regarded as false within a theory and third, it could be regarded as having an indeterminate truth value within a theory. The indeterminate truth-value is not a different kind that rivals "true" and "false" but it is rather a transient stage during which the theory has not attributed a certain truth-value to the proposition in question. When filled in, the indeterminate truth-value would still accept either "true" or "false". It is certainly conceivable that the proposition could be dropped from the theory and in this scenario the question about its truth value and meaning within the theory would not arise. No proposition within a theory, however, could afford to fail to have one of the three options; the theory should be able to semantically address every well formulated problem that falls within its scope. To a certain extent it is by virtue of its having one of the three possible truth-values relative to a theory that the proposition could be regarded as pertaining to the theory. It is part of the identification of the body of the theory that it is capable to attribute a meaning and hence, a truth-value, to every proposition that pretends to be either a part of the theory or to be addressed by it. In this sense once meaningful within a theory T the proposition P inevitably acquires one of the three values.[9]

Historical cases (like the discovery of the neutrino particle) illustrate the more difficult revisions in truth value where the scientific community takes an "indeterminate" truth value as changing to "true" within a theory. The clear-cut alternatives are perhaps even easier to illustrate. For example, the proposition that the Earth is not in the center of the universe [and it is eventually revolving around the sun] would undoubtedly be regarded as having a truth-value "false" within the Aristotle's geocentric system. After the revision of the geocentric system theory this proposition would be regarded as having received a truth-value "true". History of science shows that conjectures like this, interestingly enough, were available even before Copernicus. For example, from Aristotle we know that the Pythagoreans believed that the earth revolves around a fire that stays in the center of the system.[10] But probably the most striking illustration is the one provided by Aristarchus from Samos who, according to Archimedes, explicitly suggested that the Earth rotates around the

Sun.[11] Both versions of the proposition ("Earth rotates around the center of the universe which is a fire" and "Earth rotates around the Sun and not vice versa") would have been false when taken within the Aristotelian geocentric system. Eventually, this system suffered revision and got substituted with the heliocentric Copernican system. This in its turn could be approached as a revision of the original Aristotle's thesis that the Earth is in the center of the universe. The thesis had naturally the truth value "true" when taken within the geocentric system but received a truth value "false" as the system was modified and substituted with the heliocentric one. These simplified illustrations which address only the revision of a proposition taken in isolation and do not account for the modification of the system with respect to which the propositions are examined nevertheless illustrate the truth-value aspect of the scientific revisions.

If the above considerations are correct then the discussion of the truth value of scientific propositions is epistemically interesting even if the general context of the epistemic approach is justificatory and frames the analysis in terms of justification and not in terms of "true" or "false". This would provide one of the suggested kinds of revisions, the revision in truth value, which complements the other kind of suggested revision, the revision in meaning.

Justification and Defeasibility

How could an empirical revision render a supposedly a priori proposition not merely false but also empirical in the first place? The epistemic structure of the revision as accepted by the dominant view is roughly the following: if P is supposedly a priori justified but as it turns out P is also empirically revisable then P, if actually empirically revised, on the one hand, loses its truth value or suffers a shift in its meaning but, on the other hand, loses the epistemic kind of its justification as well. This is largely because the leading purpose of every justification is to grant or reach for a certain meaning and truth value; once the truth value and the meaning are shifted the providing justification does not manage to do the job and so P is exposed as not being justified successfully. A justification, however, is inseparable from its epistemic kind. Thus, if P is justified not merely by its justification but by an epistemic justification (which would be the conjunction of the justification and its epistemic kind) by virtue of losing its justification after the revision loses not merely the justification but the *epistemic* justification which is the whole conjunction, the justification and its epistemic kind. Therefore, P not being justified by justification J that has an epistemic kind E, loses all epistemic connection with both J and its

E. The question is: could P retain the epistemic kind E of the justification J in case the justification J, after a revision, is found not to provide good support for P? Is the revision of P's J also a revision of E? If we accept the traditional empiricist view about empirical defeasibility where an empirical defeater could revise not merely P but also its epistemic kind, the answer is yes, revision in justification goes together with revision in epistemic kind.

The underlying reasoning behind all criticism against the a priori character of some propositions' justification stems from the assumption that if a justification could be defeated empirically then it could not have been a priori. Consequently, *if a proposition is empirically fallible in case it is actually empirically revised it could not have been a priori in the first place* (A1). A closely related assumption is an assumption about defeasibility according to which *if a justification with epistemic kind E_1 could not defeat the supporting justification of a proposition, then the epistemic kind of the supporting justification could not have been E_1* (A2). So, if the justification J of a proposition P could not be defeated by empirical defeating justification chances are that J is non-empirical, that is, it is a priori. Immunity from empirical defeat is taken to be essential for justification to be a priori. The reasoning behind A1 and A2 is governed by the intuition that epistemic defeasibility is a means for epistemic kind recognition and yet this reasoning is one-sided as it stems only from the side of the empirical defeasibility.

There are two immediate problems with these assumptions. The first problem is the obvious epistemic asymmetry behind the assumptions. The argument works for the empirical revisability but not for non-empirical revisability if there happens to be one for the trivial but instructive reason that non-empirical revision as a standard is not considered. For no one seems prepared to abandon the epistemic kind "empirical" of a proposition that has been found false on non-empirical grounds and this is mainly due to the fact than no one conceives of such scenario as epistemically viable (let alone as an actual) scenario in general epistemology and especially in scientific epistemology. On the one hand this is largely due to the prevailing intuition that actually there are no a priori revisions. The main reason for this is that the epistemic kind of empirical is conceived of as having some sort of *epistemic primacy* over the epistemic kind of a priori. For if there is an epistemic symmetry and A1 holds then we should be able to formulate its mirror version (1A) which is the natural complement of A1 within the space of possibilities described by A2: *if a proposition is non-empirically fallible in case it is actually non-empirically revised, it could not have been empirical in the first place*. As a consequence, if one accepts A1 and A2 she would be able to run the mirror argument against the empiricist. Thus empiricists should devise a defense against arguments that claim that proposition p has not been empirically justified because it has been found false in a non-empirical way. In this sense all consistency problems within a scientific theory that are found problematic without any measurement, observation or experiment whatsoever would be good candidates for a non-empirical refutation of supposedly empirically justified propositions. It is not obvious at all that the empiricist would concede that the propositions revised were not empirically justified *because* they were found false a priori. And prima and secunda facie she would most probably be right. Because in cases like this it is far from clear how the epistemic kind of the defeating justification does bear on the epistemic kind of the supporting justification of the proposition at all. We do not need experience in order to know that if a proposition within a theory says that the theory is indeterministic (p) and another proposition within the same theory says that the theory is deterministic (~p) and both propositions follow from the theory itself, the theory is quite a good candidate for being an inconsistent one. If the empiricist is not prepared to accept the empirically justified propositions as a priori in virtue of their a priori revision then she does not seem, at least formally, justified to argue that [supposedly] a priori propositions lose their epistemic kinds in virtue of their empirical revision. This is dictated by the desideratum that an epistemic theory should allow at least logically for all its epistemic kinds to be on a par when it comes up to their possible relations with epistemic fundamentals like justification and revision. The next question that stems naturally but which is nevertheless a completely different question is whether all conceivable epistemic possibilities like epistemic defeat of a priori propositions due to empirical revision or epistemic defeat of empirical propositions due to a priori revision obtain in actuality; a relevant epistemic analysis of the issues within scientific epistemology and scientific epistemic practice should address this question specifically.

The second problem with the assumptions A1 and A2 is that the empiricist needs a positive story about *how* one epistemic kind could bear on another epistemic kind as it is assumed in the case with A1. For it is not the mere fact of finding p false that identifies the epistemic kind of its justification as an empirical one but it is the epistemic kind of the defeating justification that does the job. And if there is such a story it would follow at least in principle that epistemic kinds of defeating justification could identify epistemic kinds of supporting justification as the same epistemic kind and this opens a whole new door in the debate where problems like 1A are about to be examined and the epistemic battleground between justification and fallibility would never look the same.

The Misbalance between the Epistemic Kinds

Contemporary epistemology judges the two available epistemic kinds with a different standard. First and naturally all empirical knowledge is taken to be revisable. This revisability is never considered as a serious problem for its epistemic stand; on the opposite, the revisability of empirical knowledge actually contributes to its epistemic stand since it is conceived as solely empirical revisability. It seems that the satisfaction here stems from the fact that even if an empirically justified statement P has been found to be empirically false, although we, of course, abandon P we still keep the method of revision, namely, empirical revision. The intrinsic relation between the empirical justification and empirical revisability seems not only not to undermine the epistemic character of all empirical Ps but to affirm it. The situation with the other epistemic kind, non-empirical or a priori knowledge, is radically different. Traditionally a priori knowledge is taken to be inseparably bound with unrevisability. This proved to be a very useful tool in the hands of the empiricist critic: it merely suffices to show that a candidate for P is revisable, i.e. not true with necessity and that results automatically in the abandonment of nothing else but the very epistemic kind of the proposition. This is a huge and obvious *asymmetry* in the way epistemic standards treat the available epistemic kinds. This misbalance has naturally led to the dominating view that empirical knowledge is not only the sole kind of knowledge available in the sciences but also the kind of knowledge that provides the backbone and the fundamental principles in accordance with which the sciences function in the first place. Ironically, by and large this anti-rationalistic tendency has its beginning in Kant's *Critique of Pure Reason* where he ties the epistemic notion of a priori knowledge with the modal notion of necessity. However, the subsequent development in both pure and applied mathematics has shown that even the best available candidates for a priori propositions and, besides all, those of the Euclidean geometry, a paragon for unrevisable propositions, suffer from the property of being revisable.

Instead of following the perfectly conceivable track of a revisability parallel to the empirical one in cases within pure mathematics, the discipline of epistemology has ignored this option. Perhaps, this was largely due to the authority of the empirical method, which historically started playing a constitutive role which allowed it to identify certain enterprises as scientific. One way or another, epistemology followed the path of empirical revisability. This has led to encapsulating a whole epistemic domain that since remains practically unexplored. This epistemological turn, however, has at least partially contributed to what some believe to be a core problem not only of natural science but of mathematics and philosophy itself: what is the nature of the relation between pure uninterpreted mathematics and nature?[12] This question cannot be resolved without the beforehand clarification of the nature of revisability that affects both pure and applied mathematics. Even if we agree with the dominant view and accept that the epistemic nature of all the revisions in applied mathematics is empirical, this is an answer that in itself says next to nothing about the potential relation between pure mathematics and nature.

What we are left with instead is an explanatory vacuum that fails to address a legitimate epistemic domain, which deals with the epistemic character of the revisions that happen within pure mathematics and logic. The very possibility of an application consists in the options of successful and unsuccessful applications and the only regulatory mechanism that could perform a corrective function is actually the mechanism of revisability. The general framework within which this mechanism operates is the following: if a proposition or a set of propositions withstands all directed attempts for revision it is a good enough candidate for a successful application. If it does not and undergoes a revision it is a good enough candidate for an unsuccessful application. Now, without the clarification of the very mechanism of the revision the application remains stripped of criteria that would pronounce it successful or not. And the only conceived of revisability in current epistemology is the empirical one. Even *prima facie* the epistemic misbalance suggests a way out of this problem: to every epistemic kind there should, at least logically, correspond a kind of revisability. On the one hand this would balance the epistemic debate and would exhaust its niches. On the other hand a development of the conception of a priori revisability would clarify in an epistemically complete manner the possible mechanisms of revisability for both pure mathematics (and logic for that matter) and natural sciences, taken as separate as well as combined within the discipline of applied mathematics. The modal notion of necessity has to be symmetrically divorced from the epistemic notion of apriority and especially after the work of Kripke it is quite a surprise that this has not been actually done. Kitcher is a good example of a recent critic who attacks the notion of a priori through the door of necessity contributing to the general prevailing tone of the above epistemic misbalance. It is a further and completely different point whether a priori revision (except mere logical conceivability) is also epistemic and historical actuality in the case of pure mathematics, logic and natural sciences. But this is a side issue from a purely epistemic point of view. It is not a side issue from the point of view of current philosophy of science though.

Symmetry and Epistemic Kinds

The generally accepted view about the relation between revisability and epistemic kinds is that revisability is possible only within a single epistemic kind. This stems from the dominant epistemology in contemporary philosophy of science, one or another form of modern empiricism. The followers of this view intuitively accept as clear that all revisions in science have their origin and grounds in experience.[13] In the case with alleged candidates for a priori knowledge or a priori justified propositions the natural line of the empiricist answer is that no a priori proposition could continue to be accepted as a priori if it turns out to be empirically revisable. The underlying argument behind this reasoning is a synthesis of two influential theses: the *principle of fallibility*, that says that all knowledge is fallible and justified propositions even more so, and the *empirical revisability principle*, that says that in science there are no other revisions than the empirical ones. Jointly, these principles form the synthesis that provides the most common and effective epistemic onslaught against the notion of the a priori in science. In this sense, it is no surprise that there is only one possible relation between revisability and epistemic kind: there is simply no alternative to the empirical revisability conceived of and hence the two possible scenarios collapse to a single one.

Standardly, a justified proposition P is either empirically or a priori justified. If justified empirically it is considered naturally to be empirically revisable as well. The epistemic character of P is thus doubly ensured: once from the epistemic kind of the justification of P that is empirical and a second time from the epistemic kind of the revisability of P that is also empirical. If P is supposedly a priori justified it is nevertheless conceived of as still empirically revisable and because no a priori principle could stay a priori if empirically revisable P is reconsidered as being empirical in the first place even at the level of its justification and before the revision takes place.

In this way whatever epistemically happens in the contemporary epistemic domain it happens, to use a recent expression by Miscevic,[14] only within the family, and the epistemic nature of all such families in science is widely accepted to be an empirical one. The important consequence from this reasoning is that nothing could endanger the epistemic kind of all the Ps in the family: there is no epistemically outer danger that eventually would penetrate the family and taint the epistemic kind because factually there is but only one family and no neighbors inhabiting the deserted epistemic fields around. Further, no revision within the family whatsoever could change the epistemic kind of a P member of the family: the epistemic kind of all the allowed revisions is the family kind and this is a guarantee that the mechanism of any revision would end in P or set of Ps having the same initial epistemic kind as the one they had before the revision.

In this sense and especially in the case of supposedly a priori justification the prevailing view manages to keep both the justification and the revisability within its favorite epistemic kind, which has an epistemic monopoly: the empirical family. It now becomes clear that the actual function of the monopolist empirical revisability is crucial from an epistemic standpoint. First, it eliminates epistemically alternative candidates for a priori justification and second, affirms the epistemic kind of the empirically justified propositions from the opposite direction through securing the result that even after a revision they would continue to keep. The third and most important result, however, is that empirical revisability establishes the very epistemic kind as the only one actually acting; that is, even if alternative revisabilities are conceivable, they are for sure not actual, especially so in the case of sciences. In effect, the above monopoly blocks the logically perfectly conceivable possibility of non-empirical, that is, an a priori revisability.

At the core of the epistemic framework that provides the setting for the debate stays the crucial distinction between empirical and non-empirical. Without this distinction all claims of empiricism would be if true epistemically trivial. The only way to attest the epistemic nature of what is a *differentia specifica* for modern empiricism, namely, experience as the origin and the source of all justifications in sciences, is to contrast it against the possibility of its negation. Experience is experience only in the face of the real full blooded possibility of non-experience. This basic principle reveals the symmetry as a logically important and yet natural feature of the epistemic framework. Translated for the next epistemic levels of knowledge, justification, and revisability this produces the following logical space of options:

	Knowledge	Justification	Revisability
Experiential	Y	Y	Y
Non-experiential	Y	Y	Y

Table 3-1

The current picture however is quite different from this. Modern epistemology and especially scientific epistemology allows for nothing but the following:

	Knowledge	Justification	Revisability
Experiential	Y	Y	Y
Non-experiential	Science: No	Science: No	Ø
	Math: Conceivable	Math: Conceivable	Ø
	Logic: Conceivable	Logic: Conceivable	Ø

Table 3-2

While there is a clear opposition between the experiential and non-experiential rows in the case of knowledge and justification, where the points of disagreement are over the content and not over the formal possibility in the case of revisability even this opposition is not available. The reason for that is that even the logical possibility of the existence of a priori revisability is not explicitly conceived of let alone positively developed.

A priori Concepts and A priori Justification

The notion of a priori is used as holding for knowledge, justification and concepts. An important distinction, however, exists between the notion of a priori concept and the notion of a priori justification. Whereas the conceptual a priori says nothing about the way of knowing propositions built around a given concept the justificatory a priori says something about the way x is justified in holding a proposition. The conceptual a priori does not have as its main focus propositions but concepts and the justificatory a priori always pertains to propositions or sets of propositions and not to isolated concepts. The conceptual a priori is contrasted with the conceptual a posteriori: a concept is a priori if and only if it is not empirical. As a standard empirical concepts are taken to be the ones that we have or know through experience. Thus to use a well known example originally due to Frege[15] and recently discussed in an epistemic way by Miscevic,[16] I can have the concept of "WHALE" only empirically for if I did not have any cognitive contact with a state of affairs that has something to do with whales I would have probably never mastered the concept. Yet this contact, in the whale case, seems to be possible only by empirical contact and therefore my having and knowledge of the concept WHALE is plausible to be taken as empirical. The empirical nature of the concept is additionally supported by the fact that my knowing and understanding of the meaning of the concept is quite fallible for originally if I were not acquainted with the biologically relevant features of the whale I would have probably thought that to be a fish is part of what is to be a whale. Originally, whales were taken to be a kind of fish for they shared all the fish-kind features – they lived in the water, although big they look like fish, etc. Turns out whales are not fishes but mammals and this turning out became possible through experience that showed that actually whales although sharing important features with fishes are in fact mammals. What was thought in the initial concept due to some experience was found not to be the case *due to another experience*. Thus, propositions where the whale-fish concept were employed were empirically revised and the whale-fish concept was substituted with the whale-mammal concept. The revision is a classic empirical revision because the origin of knowledge that whale-mammal concept is better referring to the actual animal was experience; together with the origin of knowledge that the whale-fish concept is not so well referring. This sort of revision carries an important upshot, namely, that the subject-predicate relation in the proposition that cashes out the concept of whale as having the predicate "fish" in the initial case and as having the predicate "mammal" in the second case is both knowable and revisable through experience. This only supports the suggested empirical character of the concept; consequently, and this is an important step from an epistemic point of view, it is sometimes taken *to support the empirical epistemic nature of the propositions* where the concept figures. It is not difficult to see that this bears directly to the epistemic status not only of the conceptual a priori candidates but, what is much more important, also on the epistemic status of the propositional or justificatory a priori. In this way the opponent of the a priori elevates an argument for the empirical nature of a concept to an argument about the empirical nature of a proposition, that is, of the justification of a proposition. Clearly, this would enhance significantly the scope of the empiricist criticism against the a priori. The question, however, is whether this elevation is a legitimate epistemic move.

The rough form of the elevated conceptual argument is the following: the proposition p is not justified a priori because it employs concepts that are empirical. Again, we have to distinguish between the two notions of a priori involved, the conceptual and the justificatory. For to have or to know a concept a posteriori is to have it through experience and the experience in question does not deliver anything but the concept. For any attribution of a predicate to the concept would be forming a proposition around the concept and in this case clearly the function of the experience involved would be a different one: now it delivers not merely a concept but also *a reason to hold the proposition* that attaches a predicate to the concept. In this sense the experience delivers much more from what it delivers in the first case: it delivers a justification to hold that something is so and so. From an epistemic point of view and as far as interesting knowledge is delivered through propositions and not merely through isolated concepts and as far as justification is never a justification of a concept but a justification of a proposition this role is much more important and therefore, *qualitatively different* from the role of experience that merely delivers a concept. Is it nevertheless possible, as an objection might ask, for there to be an experience that delivers an isolated concept outside a proposition?

Is it not the case that every concept is delivered together with the propositions that affirm of it some predicate and thus every experience that delivers a concept is actually a propositional experience and thus the epistemologically interesting one? If this is the case then the empiricist elevating argument succeeds for the epistemic nature of a concept would turn out to be the same epistemic nature of the propositions where the concept functions and thus if a concept is empirical the proposition, its justification included, would be empirical as well. This objection seems to lie in the heart of the elevating argument.

Yet, it does not look to hold under scrutiny. For it seems that it is perfectly possible to form a concept empirically without at the same time forming a proposition that employs it and ascribes some predicates of it. Thus I can form a concept via crude experience without at the same time the need to be able to form a proposition about it. A child might ostensively acquire the concept DOG through her looking at animals that happen to be dogs and at the same time hearing the word "dog" from her mother pointing to the animal. With time the child will associate the sense impressions that come with observing typical dogs with the word "dog" and when prompted to show the dog she will point and say "dog". By reasonable standards it is acceptable to suppose that the child does have the concept "dog" and also manages to apply it successfully in dog cases (to apply it successfully certainly does not mean to apply it infallibly: the child might well be mistaken saying "dog" by pointing to, say, a cleverly disguised cat and this shows the fallibility of applying a concept but yet this is the same case as with whales and fish so if we accept that the whale-fish bearer of the concept is a competent bearer of the concept so should be the child; the child might need some additional experience that would discriminate between actual and fake dogs but so does the whale-fish believer and we would still prefer to accept that she is a competent possessor of the concept). Is it then essential that the child forms a proposition in order to get to have the concept "dog"? Would she fail to acquire the *concept* if she did not form a proposition like "this is a dog" or similar when first encountered with the animal or with the word? It does not seem so. For the child certainly might form a proposition while acquiring the concept but it is not necessary that she does so for she would not fail to have the concept if she fails to form the proposition. Forming the proposition requires a rational reflection over the concept and the sense data and this reflection is not necessarily present in the mere acquiring the concept. Besides all, however, it requires the coordination of the concept with other concepts and this is a pretty high order rational activity of a logically-linguistic and semantic kind. She might coordinate the word "dog" with the pointed object in front of her and this seems as good a having a concept as any other one, but this coordination is far from necessarily being a propositional one, for the child does not necessarily have to reflect over the object referred to by the word "dog" or over the features of the dog; she only needs to perceive them in order to be able to form the concept and acquire an initial understanding of the concept.

In any case, experience that leads to forming an empirical concept seems quite different from experience that leads to holding a proposition. Thus the experience necessary for acquiring a concept is by no means *sufficient* for holding a proposition that employs the concept. The child might have the concepts "dog' and "tail" but she would not be able by this experience only to be able to hold the proposition "dogs have tails". She might form the proposition by virtue of the available experience but she would not be able to hold the proposition, for if all dogs that she had seen were Dobermans with cut tails she would still have the concepts "dog" and "tail" (from a separate source like, say, a peacock) but she would not be able to attribute the predicate "has a tail" to the concept "dog" for *she does not have the experience that links the two*; not to mention all sorts of problems that come with terms like "all". It well might be the case that all dogs she has seen do have uncut tails but it is not necessarily so and in cases where she only has the experience that leads to concepts and no experience that connects them in some certain way within a proposition she would not be justified to hold the proposition even if she forms it.

If all of the above is correct then the elevating argument from conceptual epistemic kind to propositional epistemic kind fails. For if experience that brings about concepts is not sufficient to bring about propositions the epistemic kind of the concept could not bring about the epistemic kind of the justification of the proposition. Thus both conceptual a posteriori and a priori and both propositional a posteriori and a priori are distinct epistemic notions and even if there is some relation that holds in some cases among them they should not be identified on pains of commiting a category mistake.

The important epistemic consequence from this is that the empirical kind of justification of a proposition is not granted by the mere empirical kind of the concepts involved. An instructive point in relation to this has recently been made by Ken Manders when he claims that the fact that we have concepts does not mean that we have inferential means to connect them.[17] The relevant distinction between concepts and inferences is the propositionality: the latter does have this property, whereas the former does not. Also, this line of thought is more or less the spirit behind Kant's famous dictum in the Introduction to the *Critique of Pure Reason* when he clarifies the distinction between pure and empirical knowledge and argues that although all our knowledge begins with experience, it does not *follow* that it all arises out of experience.[18] Boghossian and Peacocke stress the same issue in the introduction to *New Essays on the A priori* pointing that

"In the case of a priori propositions, much experience, perhaps of a specific character, may be required to grasp the concepts implicated in the proposition or to access the entitlement to believe it; but conditions of grasp and of access remain distinct from the nature of entitlement. This is in accord with the traditional rationalist position from Leibniz onwards. Experience might be a precondition of coming to know a priori truths, but those truths nevertheless have a justification, and can also be justified for the thinker, independently of experience."[19]

Notes

---

[1] Edidin, Aron [1984] "A priori Knowledge for fallibilists" in *Philosophical Studies*, 46, pp.189-97, p. 189.

[2] The most full exposition of Brown's original position is given in Brown, James [1993] *The Laboratory of the Mind: Thought Experiments in the Natural Sciences*, Routledge, New York.

[3] For details see Norton, John [1985] "Thought Experiments in Einstein's Work," in *Thought Experiments In Science and Philosophy*, eds. T. Horowitz, G. J. Massey, Savage, MD: Rowman and Littlefield, and Norton, John [1996] "Are Thought Experiments Just What You Thought?" in *Canadian Journal of Philosophy*, 26, pp. 333-66. For Norton's most recent work on thought experiments see also Norton, John [2002] "Why Thought Experiments Do Not Transcend Empiricism" pp. 44-66 in Christopher Hitchcock (ed.) *Contemporary Debates in the Philosophy of Science*. Blackwell, and Norton, John [2004] "On Thought Experiments: Is There More to the Argument?", Proceedings of the 2002 Biennial Meeting of the PSA, *Philosophy of Science*, 71, pp. 1139-1151.

[4] Thurow, J. [2006] "Experientially Defeasible A priori justification' in *The Philosophical Quarterly*, Vol. 56, No. 225, pp. 596 – 602, p. 596.

[5] Kitcher, Philip [1983] *The Nature of Mathematical Knowledge*. New York: Oxford University Press, p. 24

[6] Grice, H.; Strawson, P. [1956] "In Defense of a Dogma" in *The Philosophical Review*, Vol. 65, N. 2, p. 157.

[7] Casullo, Albert [2003] *A priori Justification*, Oxford University Press, NY, p. 10.

[8] Friedman, Michael [2001] *Dynamics of Reason: The 1999 Kant lectures*, CSLI Publications, Stanford, p. 83-91.

[9] For example, within the standard model of contemporary quantum mechanics the proposition "There exist a physical object x (be it particle, wave or both or something else) with such and such properties that is called "neutrino"" is regarded as having a determinate, though relative to the theory, truth-value "true". This, however, was not the case until the actual discovery of the neutrino in the fifties of the previous century by Fred Reines and Clyde Cowan (For details see C.L. Cowan, Jr., F. Reines, F.B. Harrison, H.W. Kruse and A.D. McGuire [1956] "*Detection of the Free Neutrino: A Confirmation*, in Science 124, 103; Frederick Reines and Clyde L. Cowan, Jr., [1956] "The Neutrino", Nature 178, 446 and

Frederick Reines and Clyde L. Cowan, Jr., [1957] "Neutrino Physics" in *Physics Today* 10, no. 8, p.12.) Until then the neutrino was only a hypothetical object, baptized so by Enrico Fermi in 1933 who gave a name to what already was postulated by Wolfgang Pauli in 1930 as a solution to the problem of the missing energy in the nuclear beta decays. Whatever the truth value of the above proposition before the actual experimental detection of the object, it was not the same as the truth-value after the detection. The hypothesis received experimental confirmation besides the synchronously developed theoretical body that made extensive use of the postulation and after the experimental confirmation we have good reasons to believe that the proposition acquired the truth-value "true" within the theory that employed it. Now, if the truth value before the detection was different from "true", because truth in contemporary physics is generally regarded as unachieved with scientific certainty given the absence of experimental confirmation, the truth-value of the proposition between the thirties and the fifties of the previous century must have been different from "true". The scenario where the proposition has been "false" until the experimental discovery does not seem very attractive – no reason was available to regard it as *not being true* before, let alone after, the detection. After the discovery the very proposition did not change; what has changed is the body of the theory that now employed the new proposition or set of propositions that proved the physical existence of the neutrino experimentally. Also, it is not plausible to accept that the proposition did not have a truth-value at all before the detection, for it was embedded within the theory. By elimination the scenario where the proposition was considered to have an indeterminate truth-value until the discovery seems like the most promising option left.

[10] Aristotle, [1992] *Physics*, Books I-II, translated with introduction and notes by William Charlton, Oxford: Clarendon Press (Clarendon Aristotle Series), 13; 293 a 19.

[11] For the Archimedes work *The Sand Reckoner* see Heath, Thomas L. (TRN), [2002] *The Works of Archimedes*, Courier Dover Publications, NY, pp. 221 – 233.

[12] The problem was raised famously by Hilbert a century ago and with few exceptions (like Eugene Wigner and recently Mark Steiner) it is consistently neglected ever since

[13] *Locus classicus* for the last century or so is Willard Quine with the famous dictum that all statements face the tribunal of experience (TDOE in *The Philosophical Review* 60 (1951): 20-43. This line continues forcefully even at present day. Modern empiricist put aside, even rationalists as Bonjour try to avoid the problem of empirical revisability and do not argue against it directly (Bonjour, L. [1998] *In Defense of Pure Reason*, Cambridge University Press, UK, p. 10, 15 – 20, 120 - 124). In his recent and impressive survey "A priori justification" Casullo (Casullo, A. [2003] *A priori Justification*, Oxford University Press, NY) symptomatically for the prevailing epistemic attitude argues that the only way we could know what a priori justification is and whether actually there is one is to pursue an empirical investigation; no positive story is provided for the conceivablity of a priori revisability. Finally, Michael Friedman in *Dynamics of*

Reason tries to show how an apriorist might actually evade the problem with the epistemic revisability.

[14] Miščević, Nenad [2005] *The Three Grades of Immunity*, unpublished lectures, CEU, Budapest

[15] Frege, Gottlob [1885] *Foundations of Arithmetic*, Oxford, Blackwell.

[16] Miščević, Nenad [2001] "Apriority and Conceptual Kinematics" in *Croatian Journal of Philosophy* (1 (1)/2001), pp. 21-48.

[17] I am indebted to Ken Manders for clarifying this point in his lectures on philosophy of mathematics at Pittsburgh University during the fall of 2006.

[18] Kant, Immanuel [1781/87] *Critique of Pure Reason*. Trans. Norman Kemp Smith as *Immanuel Kant's Critique of Pure Reason*. London: Macmillan Co. Ltd., 1963, Introduction.

[19] Boghossian and Peacocke (eds.) [2001] *New Essays on the A priori*, OUP; Introduction, p. 2.

CHAPTER IV

A PRIORI REVISABILITY IN NATURAL SCIENCE

The chapter suggests and positively develops the notion of a priori revisability for natural science. It starts by delineating some relevant background from the debates on the nature of a priori that bear on the problem of a priori revisability. Further, it continues by demonstrating the epistemic difference in kind between paradigmatic empirical knowledge and prominent candidates for a priori knowledge (like mathematical ones). Accepting both universal fallibility and epistemic symmetry it forms the notion of a priori revisability in contrast to the dominant notion of empirical revisability and defends it epistemically. Some relevant objections are discussed. Section one sketches the relevant historical background. Section two argues for the conceivability of the a priori given the principle of fallibility. Section three discusses the modern concept of the a priori. Section four suggests the notion of a priori revisability. Section five examines and responds to the problem of epistemic homogeneity. Section six addresses the problem of the uniqueness of a priori revisions.

Some Historical Background

The notion of a priori has been in the focus of many of the epistemic discussions throughout the twentieth century. The logical positivists tried to salvage in their own original way what was left from the Kantian rational program. Reichenbach[1] and Carnap[2] attempted to defend the notion on semantic grounds; they argued that analytic propositions are a priori and since there are analytic propositions then there are a priori propositions as well. This sort of defense, albeit being too far from Kant's original strive for synthetic a priori propositions in order to be called truly "Kantian", is exemplary for much of the philosophical strategies to not only defend the a priori but also to attack it. The key common feature in those strategies is to step on *non-epistemic* grounds, like modal or semantic, and to try to come up with a conclusion that concerns the epistemic notion of a priori. Historically, these kinds of arguments stem from Kant's own original position where he tied the a priori with the modal notion of necessity and with the semantic notion of analyticity.[3] Thus, it is no wonder that arguments concerning the same subject matter followed in the same steps. After the logical positivists, in 1963, Quine famously attacked the distinction between analytic and synthetic propositions and many see in this an effective attack against the notion of the a priori as well. Again, the underlying reasoning behind this interpretation is non-epistemic: the notion has been attacked indirectly through the semantic notion of analyticity and many philosophers took it to be successful for they were convinced in the intimate relation between the analytic and the a priori.

Great portion of the debates on the a priori after Quine discussed the notion mainly through the looking glass of analyticity and revisability. Quine took a priori propositions to be rationally unrevisable and since he held that "no statement is immune from revision"[4] this automatically led to the thesis of the non-existence of a priori knowledge. In his first indirect attack on the a priori, based on analyticity, Quine accepted an intimacy between the analytic and the a priori of the highest grade. In his second, direct, attack he accepted similar intimacy between the a priori and the necessary. In this way the two perhaps most influential modern criticisms of the notion of the a priori are non-epistemic: they do not argue against the a priori on purely epistemic grounds but only through alleged or even assumed relations between the a priori and the notions of analyticity and necessity. This trend continued for quite a while after Quine and thus in 1983, not a very good year for the a priori, Hilary Putnam defended the position that the notion of a priori presupposes rational unrevisability.[5] During the same year Philip Kitcher argued that the notion of a priori implies the notion of necessity.[6] When taken together with Pierce's influential principle of fallibility (or fallibilism)[7] that every proposition might be revised and consequently, that might turn out to be false, these imply that a priori knowledge does not exist. Recent work, however, showed that all of the non-epistemic attacks suffer from an old and, ironically, Kantian dogma, namely the view that there is indeed a strict dependence between the analytic and the a priori, on the one hand, and the necessary and the a priori on the other. However, in 1972, Saul Kripke[8] and more recently Albert Casullo (in the eighties and most recently in 2003)[9] demonstrated that this dogma should be abandoned for the distinctions involved are independent. Kripke argued that there are contingent propositions that we could know a priori as well as necessary propositions that we could know a posteriori and Casullo argued about the independence based on their difference in philosophical kind. Recently, Nenad Miscevic offered an original argument that (even) analyticity does not entail apriority.[10] Today most contemporary epistemologists agree that the analytic-synthetic distinction is a logical and semantic one, that the necessary-contingent distinction is a modal and metaphysical one and that the a priori – a posteriori distinction is an epistemic one. Thus to argue against one term of a distinction as if it is coextensive or even partially coextensive with another term from a different distinction is a kind of category mistake and especially so if no explicit argument for the alleged co-extensivity is offered. To accept the independence of distinctions is certainly by no means to deny that a relation between the notions exists. To argue, however, that there is some sort of relation between the terms is very different from *assuming* that there is such a relation. And the former is naturally much more difficult from the latter. In fact no uncontroversially accepted relation between the terms of the three relations is available in current epistemology and instructively, the nature of these relations has inspired some of the most interesting work in the field.[11] Thus the old Kantian view that the notion of a priori implies necessity and that the notion of analyticity implies apriority have to be abandoned until the opposite is proven explicitly.

Historically, the most influential attacks against the a priori have been non-epistemic. Both kinds, the semantic attack through analyticity and the modal attack through necessity rest on a common notion, the notion of revisability. Analytic propositions are considered a priori to a great extent because we cannot conceive of a possible denial of the predicate relation to the subject and thus we appeal to a modal property of the proposition, namely, its inconceivable revisability in both meaning and truth value. Necessary propositions are traditionally considered as prominent a priori candidates because of their unrevisability too. In both general spectra of cases revisability plays an important role in the identification of a particular case of knowledge as a priori. The significance of revisability of propositions has increased considerably in recent years when epistemic analyses tend to be formulated more often as analyses of epistemic justification and not as analyses of knowledge. For it is predominantly accepted that no justification guarantees knowledge with certainty and thus the question of its strength naturally emerges. Both strong and weak justifications might be examined and this examination amounts to revision when evidence is found that modifies the success of the justification. The defeasibility of epistemic justification practically establishes the principle of fallibility as the perhaps most uncontroversially accepted principle in contemporary epistemology. Being fallible for a proposition means that it is revisable at least in principle and since every revision is to be conducted through some sort of additional epistemic justification propositions thus become *epistemically revisable*. The epistemic revisability is identified through the establishing of which epistemic kind, the a priori or the a posteriori, enables it. In this way the question of the epistemic revisability bears directly to the question of the epistemic nature of propositions.

Revisability and the Conceivability of A priori

Prima facie, it seems almost inconceivable for a proposition that is revisable, and according to the principle of fallibility all propositions are revisable, not to be revisable empirically. In fact, this seems to be the prevailing view in current epistemology. Much of the force behind this alleged inconceivability stems from the reigning supporting epistemology, one or another form of contemporary empiricism. In the light of the view that most if not all of our knowledge comes from experience it is most natural to suppose, and often perhaps tacitly, that the regulatory mechanism that tests the knowledge is *also* empirical by nature. In addition, this view is complemented with the negative view that even if alternative epistemic sources and routes of justification are conceivable they are to be too vague and ill formulated to be useful. Yet the dominant view is again far from uncontroversial. For it is notorious that empirical knowledge is to a great extent contingent upon many a condition, not at all in the controlling power of the cognitive agent. Thus perceptual knowledge, apart from exotic cases of illusions, hallucinations and mirages, is heavily dependent on the actual mechanism of perception and it is well known that this mechanism, and not just because of its complexity, is very difficult to be accepted as delivering infallible knowledge. But perhaps the worst part is the mysterious relation between the more or less well understood physical component of the perception and the rational formation of concepts and propositions. Thus, to grant confidence to the predominant source of our knowledge is one thing and to argue that alternatives are inconceivable is quite different. Fallibility of empirical knowledge is at best compatible with its trustworthy character but cannot demonstrate its uniqueness. For clearly, there are cases where we would be quite reluctant to deny that we do indeed have knowledge and yet this knowledge does not seem to come from experience. Mathematics and logic provide the usual illustrations.

Knowing the exact weight of a given rock is so obviously dependent on our observation and measurement abilities that it is almost impossible rationally to deny that it is a piece of empirical knowledge. We need to have our senses, aided or not, to be in a causal contact with the precise object of our interest under some specific conditions (like standard gravitation (the rock should not be measured, say, as it floats freely within a plane heading super fast towards earth), the object should be motionless, no other forces should be present that could influence the measurement, etc.) and for quite some while so. Then, having acquired sense information we somehow process it and form the proposition "This rock weights 287 kilograms" having the measurement conditions in mind. Also, it is almost inconceivable what else but experience of the same kind could be able to undermine this proposition, whether its meaning or its truth-value. For whoever wants to doubt it seems that must be in the same sort of contact with the rock we were, when we were forming the proposition. For otherwise, among other things, it is not at all clear how could we be certain that her undermining proposition would concern *this* rock precisely and its properties precisely, weight in particular, and not others'. Yet, this seems in the same obvious way *not* to be the case when we hold a mathematical proposition like "6851037 times 7187 equals 49238402919". For we still have our senses, perfectly tuned and yet it is obvious that in the case of forming and subscribing to the proposition we are not in contact with anything that would even remotely remind us the rock scenario. Or, to give most charitable reading, even if we somehow are indeed in empirical contact with what makes the mathematical proposition true (and in its own right this is too problematic; for whatever makes true mathematical propositions it does not seem to be of the same kind as whatever makes true empirically delivered propositions and in the literature most agree that it is not causally accessible) or, if we *were* in such contact some time before and somehow this contact delivered the knowledge encoded in the mathematical proposition, it would not be the *same* contact as in the case of the rock observation. Do we have knowledge delivered by this proposition and myriads others of its kind? We better do. Do we actually observe or measure something, whatever it is, that leads us to forming and holding this proposition? It does not seem so. A blind born person not in ordinary sort of sense contact with anything numerable would be able to rationally form the proposition and to be justified about holding it. Thus it seems that we have good enough reasons to accept that the two cases of knowledge, the rock case and the number case, are of a *different kind*. Epistemically, this is as interesting as it gets.

Perhaps the most traditional response to mathematical examples like this follow empiricism about mathematics line. First, empiricists argue that even if we did not actually form a proposition of this kind counting pebbles and learning somehow about their mathematical properties (John Stuart Mill famously held that numbers are properties of objects[12] and Frege, also famously, heavily criticized him for this in his seminal *The Foundations of Arithmetic*[13]; recently Philip Kitcher offered a view[14] that is reminiscent of Mill's empiricism about mathematics; Miscevic's position, the way I see it, is not very different in spirit, too) we did this in the past and this is why now we are able to rationally cope with cases where counting would not be of much help. The main idea is that we acquire from experience simple foundational knowledge about mathematics as well as the rules for its manipulation and thus become able to reach knowledge for cases where observation seems hopeless. In this view the knowledge, delivered by mathematical propositions of the above sort is ultimately empirical. There are many problems with such view and in the case of geometry I argue against it at greater length in chapter V. A few points, however, seem relevant here. First, the Millean metaphysical claim about numerosity as a sort of intrinsic property of physical objects and thus observable with the aid of ordinary perception is almost universally rejected. One need not be a Fregean about mathematics in order to see the appeal in his arguments against Mill's view on mathematics; it is, however, beyond our present purposes to comment on these arguments. Second, empiricism about arithmetic perhaps turns out to be even more difficult to defend than empiricism about geometry. Whereas in the case in geometry many empiricist philosophers of mathematics see it as obvious for the observation to deliver the truths of geometry by appeal to the observability of geometrical properties of physical objects. Thus, they argue for causal contact between the cognitive agent and the geometrical properties of say, being straight, linear, circular or rectangular. I will not address here the confusion between pure and applied geometry which usually underlies this view (for a discussion on this problem see chapter V and especially section 4), but I will use it to contrast with the case of arithmetic. Finally, to appeal to physical events of counting and the like and extrapolate to knowledge is to confuse forming a proposition with having reasons to hold it. I can surely form the proposition $7 + 5 = 12$ via counting pebbles but the experience in the case contributes to my forming the proposition and not to its holding it true; for the proposition is about numbers and I certainly do not count numbers as I count pebbles.[15] Pebble counting might be *instructive* about the confidence we have in applied mathematical cases for simple situations but is epistemically far from sufficient to deliver reasons to epistemically uphold a purely mathematical proposition. And propositions of pure mathematics are obviously prior to propositions of applied mathematics and the ones the epistemologist actually cares about at the end of the day.

Further, the empiricist approach in geometry does not seem to apply in the case of arithmetic. And given the universal reduction of geometry to arithmetic[16] this does not seem to help empiricism about *mathematics*. For while in the case of geometry the empiricist had some candidates that were causally accessible by the cognitive agent, she does not seem to have ones in the case of arithmetic. It is more or less understandable when the empiricist argues that she formed a proposition about the triangularity of a table by merely observing the table and its shape. But it does not seem that she formed the proposition "6851037 times 7187 equals 49238402919" by observing anything of the kind whatsoever. The proposition "This table has a triangular form" seems to be made true by nothing else but the table and its having a shape of the mentioned kind. The main reason for that is that the proposition is *about* the table and *about* its shape. If something that is *not about* the table and *not about* its shape made the proposition true that would be of great surprise for the empiricist in the first place. Thus, it seems that what we call referents of the singular terms of the proposition (the actual physical object "table") and their properties (the actual physical property "triangularity" of the physical object "table") are what the cognitive agent needs to be in contact with in order to reach a true proposition about them. Yet this is far from being the case in arithmetical cases. The proposition "6851037 times 7187 equals 49238402919" *is not about* tables and their shapes. It is about numbers. In fact it is about particular set of numbers, "6851037", "7187" and "49238402919". It is valid only about them and about nothing else. For the terms that constitute it are about those particular numbers. And numbers, those in the proposition or in any other arithmetical proposition, are not available for empirical observation. Whatever they turn out to be they are not something that could be looked upon and thus something that could empirically demonstrate its properties. Therefore, they differ not only from clear-cut cases like the rock case but also by middle-ground, so to speak, cases like geometrical ones. On pains of denying that the proposition is *about* numbers the empiricist should concede that they are of the same semantic form as geometrical and physical propositions. And to deny that the proposition is about numbers would not be sufficient. For the empiricist would need to deny their very "about" attitude. And this would be too much on virtually any account. For if mathematical propositions are not *about* something, whatever it is, this amounts not only to their not being meaningful in the same way as the other propositions, but also to their not being meaningful in the most natural way. For terms are inherently intentional and this is where their meaning stems from. Propositions build up upon terms. The English word "table" is about a table, whether there is a table or not. The notorious term "unicorn" is about unicorn whether unicorns physically exist or not. The worst, perhaps, of all examples is the famous "round square" term. Whatever semantical nightmare turns out to regulate it the term *is about* the object "square" which has the charming in the case property of being also "round". The actual nature of the object and its ontological problems are completely irrelevant to the fact that the term is *about* this object. The aboutness is a an intrinsic feature of language, natural or formal, and to deny it to some terms of the language and allow it for others would be most unnatural from both philosophical and linguistic points of view. Human rationality works heavily through the linguistic attitude of "being about" and the burden of the proof for an opponent of this view is on her and not in an easy way. Thus causality, the main tool of empirical knowledge, fails to serve in domains like the mathematical one. The referents of the terms of the mathematical propositions are anything but causal and thus experience could teach us virtually nothing about them.

The reasoning from the "aboutness" as a crucial property of (well formed) propositions is in harmony with both the prevailing scientific semantics of a Tarskian type, as well as with prevailing reasoning in mathematics. Recently, Bob Hale stressed both aboutness and the important role of the nature of the entities that the propositions are about in the context of mathematical knowledge:

> "What is called for is a philosophical account of how we know, or what entitles us to accept, the mathematical theories we do accept. Since such an account cannot very well be attempted without adopting some view about the nature of the entities of which the mathematical theories treat, this is likely to involve broadly metaphysical questions as well as epistemological ones."[17]

To come back to the question from the beginning of the section, true or not, it therefore seems quite conceivable for a proposition to be delivered independently from experience. But the important point is that although delivered independently from experience, this proposition is still perfectly fallible. For the route of propositional delivery is only a route and not a king's road to truth. Thus, if revisability is just another function of the epistemic justification it is also conceivable that a priori justified propositions are a priori revisable.

The Concept of the A priori

The nature of the a priori notion has been approached in the literature both epistemically and non-epistemically. The characteristic of the epistemic approaches proper is to provide exposition of the a priori concept without appeal to non-epistemic notions like the semantic notion of analyticity or the modal notion of necessity. The characteristic of the non-epistemic approaches is to provide exposition of the concept through an appeal to these notions. I follow recent work by Casullo[18] and take that the epistemic approach is grossly the better one. I will not dwell here on the details of the reasons Casullo stresses but I will provide a brief one of my own. Clarifying that the proposition "There are 9 planets in the Solar system" is such that it might turn out not to be true, that is, to be a contingent proposition, says nothing whatsoever about the actual epistemic route of the cognitive subject to actually hold it. In reverse, the fact that the proposition "937362 times 62728 equals 58798843536" might turn out to be false says nothing about the actual epistemic route of the cognitive agent to hold it. Also, even if a subject of a proposition might be successfully analyzed in terms of predicates it "contains" this again says nothing about the way the cognizer came to hold it, although might be, quite in some cases, instructive about it. Therefore, I will attempt no non-epistemic exposition of the concept of the a priori at all. In addition, again partly following Casullo, I will not attempt comments on non-epistemic definitions of the a priori, like, for example, the recent one by Philip Kitcher.[19] I regard such definitions as bordering on being irrelevant to the epistemic nature of the a priori and thus as not particularly helpful.

History of philosophy shows two major epistemic lines of defining the notion of a priori. The first is the *negative line*, where a priori is being defined by saying what it is not and the second is the *positive line* where properties of the a priori are prescribed explicitly. The reality is that there is no single generally agreed upon definition of the a priori and this is the source of much of the controversies around the nature of the notion. Yet, there exists a standard definition which is at the core of the idea and this is the definition of a priori as independent from experience; most people agree that this definition captures the idea sufficiently well. Where they disagree is the actual delineation of the terms "experience" and "independence". Obviously, the relevant senses of experience and independence have to be spelled out before a piece of knowledge is to be examined as being a priori or not. In the case with experience most agree that sense experience should naturally be part of the definition and some agree that memory should be included as well. Much fewer people agree that introspection is to be part of the relevant sense of experience. The balance is really a delicate one: too broad a definition of experience would render pretty much everything a posteriori in an uninteresting way and too narrow a definition would allow for too much a priori too easily. Nevertheless, as far as much of the philosophical interest behind the notion of a priori stems from its contrast with the uncontroversial standard experience, that is, propositions delivered either directly through perceptual experience or traceable back to such propositions, there exists a sense in which the notion is epistemically sufficiently interesting when contrasted with core paradigmatic cases of experience. In this way and especially for the purposes of the scientific subspecies of knowledge and justification, it should suffice to restrict the relevant sense of experience to direct standard and technologically enhanced perceptual experience and propositions actually traceable to such experience. A proposition contrasted with such experiential dependence is perhaps sufficiently interesting from an epistemic point of view in order to be classified as an a priori one.

The sense of the independence is the true battleground for the a priori. Spelling out the nature of the independence would provide an epistemologist with precise means to assess candidates for a priori. It is this precision which, however, misleads us the most and sometimes even leads to outright rejection of the a priori as being too vague a notion and ill defined because the notion of independence from experience is incomprehensible. Yet, there are two general kinds of independence that are immediately attractive. The first one concerns the justification of the proposition: if the justification is not empirical and does not rest in an identifiable way on empirically traceable considerations then it is independent from experience. On this scenario the justified proposition is considered to be a priori justified. The second one concerns the revisability of the proposition and comes in two flavors. The strong sense is the general revisability: if a proposition is not revisable then it is independent from experience. The weak sense is the empirical revisability: if a proposition is not empirically revisable then it is independent from experience.[20] Quine, Putnam and Kitcher all opted for the strong sense of revisability where the mere fact of revisability is taken to establish an epistemic property of the given proposition. This, as I argued above, is quite controversial for both reasons of independence of distinctions as well as for problems with arguments that purport to establish the relation between the a priori and necessary as intimate enough. The most obvious problem with the strong sense of independence, however, is a different one. Affirming something about the possibility of a proposition to be examined with respect to its meaning and its truth value says absolutely nothing about the epistemic way the cognitive agent justifies herself in subscribing to the proposition. In this way general revisability of a proposition is hopelessly irrelevant to identifying the epistemic nature of the acting justification. Recently, a similar point has been made by Casullo.[21]

It is more or less uncontroversial that the justificatory independence of a proposition from experience is a necessary condition for its being a priori. It is even more uncontroversial that it is not a sufficient condition. The view that if a proposition is not revisably independent from experience then it is not a priori is a very influential one and not merely in its strong generally epistemic sense. Even if we agree that the general revisability independence of a proposition is epistemically irrelevant for its epistemic kind the epistemic revisability independence remains in power. The empiricist proponent might argue that p is not independent from experience because it is empirically revisable and this clearly seems like a kind of dependence on experience. The point of the empirical revisability argument against the a priori however should be appreciated in all its subtlety. For as Edidin demonstrated in 1984[22] and most recently Casullo (in 2003 building up on his work from 1988),[23] the argument that proposition is not a priori because it is empirically revised is seriously flawed. They argue (independently) that to pronounce a proposition which is allegedly justified independently from experience an a posteriori one merely because it is empirically revisable is to confuse supporting evidence with defeating evidence. Or even worse, to presuppose tacitly some regulatory mechanism between the epistemic kind of the supporting evidence and the epistemic kind of the defeating evidence according to which the former loses its kind due to the kind of the latter. In chapter 2 I have argued extensively against such mechanism. Also, I am more than happy to follow Edidin and Casullo in their thesis about confusing defeating evidence with supporting evidence. It is important however, to stress that the reasoning behind this confusion is at least technically perhaps due to the modern conceptual framing of epistemic problems in terms of justification and not knowledge. I have earlier offered an argument against the view that defeating epistemic kind could influence the epistemic kind of the supporting evidence or justification. There is a danger, however, if we follow this line too narrowly. For one might accept that the only sense in which we should understand the a priori is then as pertaining solely to the justification of the proposition and thus that there is no sense in referring to propositions as being a priori rather than just referring to their justification being a priori. And this is far from obvious. For the notion of a proposition known independently from experience is a perfectly legitimate notion and by no means identical to the notion of a proposition being justified independently from experience. Justification is surely a necessary condition about knowledge but it is not a sufficient one, and thus there is more to attributing the predicate "a priori" as being had in some way by the proposition than its mere attachment to the knowledge component of justification. To illustrate, if revisability of a proposition is indeed a regulator of its delivering knowledge, then being merely a priori justified would not amount to the proposition being a priori known. Therefore, it seems that Edidin's and Casullo's as well as my argumentation against the bearing of epistemic kinds leave something out. And this is the overall epistemic character of the proposition which could well be referred to as having its own epistemic kind.

Edidin's and Casullo's arguments deprive the empiricist criticism of one of its main weapons: to argue that a proposition is not a priori justified if it is empirically revisable. This does not manage, however, to deprive the empiricist criticism of its more general weapons: to argue that even when a proposition is justified a priori this does not amount to its overall epistemic character as a priori if the proposition is empirically revisable. For to accept that a proposition retains the epistemic kind of its justification whatever the epistemic kind of its defeater is one thing and to accept that a proposition is independent from experience *merely due to the epistemic kind* of its justification is another thing. Clearly, the apriorist could not stay content with the first option. For if her a priori justified proposition turns out to be empirically revisable or revised this would undoubtedly reduce the independence from experience of the proposition to mere justificatory independence. As a result, the proposition would not be generally epistemically a priori for it is revisably dependent on experience. The apriorist needs to do better in order to secure the overall a priori kind of the proposition. She needs to show that it is also revisably a priori. In this way the proposition would be both justificatory and revisably a priori and in this way the epistemic kind would cover both notions of justification and revisability thus addressing directly the epistemic role of revisability. And as we have seen above, revisability has been the door for most of the attacks against the a priori. Thus a new, positive criterion about the a priori should be adopted: to be (among other things) a priori revisable. This criterion joins the larger criterion of independence from experience and adds significantly to the available precision for identifying epistemic candidates as a priori. The criterion, however, comes with a price for the apriorist. For now she has to demonstrate not only that the justification for p is independent from experience but that also p is revisable independently from experience. And certainly this is much more difficult to achieve. In this way perhaps great number of a priori candidates would be eliminated as not being overall a priori by virtue of their being empirically revisable. But on the other hand were there cases of a priori justified propositions that are also being shown to be a priori revisable it would be much more difficult for the empiricist to argue that they are dependent on experience. For justification and revisability, the two influential sources of independence, would be both covered.

The Notion of A priori Revisability

Rational revisability was famously at the core of the debates on the a priori for many years. The rationality here, however, is not to be confused as coming from the rationalist camp. It is just another term to refer to revisability as being justified, that is, as the cognitive agents performing the revision having rational and epistemic reasons to conduct it. The conceived of epistemic nature of the rational revisability has been nevertheless strongly biased. For virtually no epistemic alternative to empirical revisability was discussed. The main reason for this perhaps is the trend to philosophy being modeled on science, and since science was considered obviously empirical, its empirical nature was naturally to be found in the scientifically modeled philosophy as well. Exceptions to the rule started to emerge relatively late. Edidin conceived of the possibility of non-empirical defeating evidence[24] in 1984 and later, in 1988 Casullo added to the vitality in the notion.[25] Most recently Thurow addressed the possibility as well.[26] None of them however, explicitly defended the notion as something transcending mere conceivability. On the opposite, they all offered arguments against the a priori. Edidin and Thurow argued for the vulnerability of the a priori due to testimonial empirical revisability and Casullo's stand against the a priori might be viewed even as a reliabilist one. Thus, at best, the notion of a priori revisability is taken as conceived of but not much more than that. And clearly, this is far from enough for both the apriorist and the empiricist. Since the apriorist needs a full-blooded notion supported with actual illustrations and the empiricist needs the same in order to criticize it.

As we saw in chapter two a proposition is empirically revisable if the epistemic kind of the defeating justification is empirical. This simple and perhaps uncontroversial scheme reveals the acting general epistemic scheme:

1. [Epistemic Justification that P; P]
2. [Epistemic Justification that Q; Q]
3. [Q is incompatible with P]
4. [Epistemic justification that Q stronger than epistemic justification that P]
5. [P is revised]
6. [Epistemic justification that P*; or that R]

In the predominant cases of empirical revisability the epistemic kind in step (6) is "empirical" because the epistemic kinds in steps (1), (2) and (4) are "empirical". The a priori revisability would be thus expressed, following epistemic symmetry between epistemic kinds in

1. [A priori Justification that P; P]
2. [A priori Justification that Q; Q]
3. [Q is incompatible with P]
4. [A priori justification that Q stronger than A priori justification that P]
5. [P is revised]
6. [A priori justification that P*; or that R]

Now, an objection seems relevant. If revisability independence from experience is indeed a part of the concept of the a priori is it not the case that an attempt to define a proposition as a priori through appeal to its revisability independence from experience is begging the question? For it seems that we should not be allowed to use revisability independence from experience as a priori in the first place since we still do not have the concept of the a priori and consequently, an attempt to define the a priori by assuming the notion of a priori itself is circular. Well, not really. For we do not try to define the concept of a priori per se by appeal to the notion of revisably a priori but only the overall epistemic kind of the proposition in question. We are already working with a general definition of a priori at hand and this is the standard definition of a priori as independent from experience. This general definition might be applied to at least three bearers (if we ignore for the moment its application to concepts): to justification, to revisability and to overall epistemic kind which consists in the epistemic sub-kinds of justification and revisability. Thus we use the revisability a priori notion in order to establish an overall a priori notion and not to define the concept.

The acting component of the revision is a proposition or a set of propositions that concern the subject matter of the revisable proposition P. Roughly, we could distinguish between several revisability scenarios. First, the revisionary propositions, although concerning the subject matter of P, do not manage to modify either its meaning or its truth-value. Second, the revisionary propositions might modify the meaning of P without modifying its truth-value; usually these are minor adjustments and shifts in the meaning that are not sufficient to change significantly P so that we start to regard it as a different proposition altogether. Third, the revisionary propositions might modify both the meaning of P and its truth-value. Last, they could revise just the truth-value of P and not modify its meaning. These are only logically conceivable scenarios and not necessarily actual ones. For example, it is not obviously clear that revision in truth-value could leave the meaning of the proposition intact; that depends on the employed semantics of the proposition. Also, it is not unconditionally clear that even slight modification of the meaning of P could refrain from change in truth-value. For what is worth, one might argue that even the slightest change or shift in the meaning of P leads automatically to the birth of a new proposition, P* or Q and this new propositions comes with its own new truth-value. The truth-value might coincide with the truth-value of P but that does not mean that it is the same, for it is a truth-value of a different proposition. Again, this depends on the semantics that regulates the meaning and the truth-value of P and its related proposition. But in the general case and especially in mathematics and natural sciences, a proposition is usually taken and uniquely identified within a system of propositions. Thus any of the axioms of geometry and arithmetic has a fixed place in their systems and even if fluctuations in its associated meaning are present usually this is not sufficient to result in regarding the proposition as a new one. The same holds for the truth-value. Often and especially in natural sciences propositions suffer from shifts in meaning. As a rule this is usually not sufficient to regard them as having a new truth-value *because* they are new propositions.

In all those cases the epistemic job is done not just by the proposition but by its supporting justification as well. In this way there is no epistemic difference between the supporting justification and the revising one besides the obvious difference in function. Thus both kinds of justifications have the same epistemic aspect; they are described epistemically as having certain epistemic kind. In order for a revision procedure to be a priori the epistemic kind of the revising proposition has to be independent from experience. In order for the overall epistemic kind of the proposition to be a priori the epistemic kinds of both the supporting and the defeating justification have to be independent from experience. The natural place to look for a priori revisions are the traditional a priori domains such as the domains of the pure mathematics and logic. Obviously, as history shows us, both domains evolve and develop and not all the time cumulatively. The contrast with the development of the natural sciences is certainly a great one: the percentage of natural scientific propositions that are abandoned as false is much greater than the abandoned false propositions in mathematics. Most of the time false propositions in mathematics are due to calculation or reasoning mistakes whereas most of the mistakes in natural sciences come from the wrongly interpreted empirical data or unsuccessful theory. In one way or another, both mathematics and natural sciences are capable of finding (some, as it were) of these mistakes and correcting them. The standard view in science is that new empirical data can correct old false hypotheses whereas in pure mathematics this does not seem to be the case. Mathematical knowledge seems to be of a different kind than the natural scientific knowledge and in this way it is natural to suppose that not only the supporting justificatory procedures, which are by and large logical procedures, but also the correcting procedures are of the same kind. At the end of the day mathematical knowledge is the end product of justification of propositions examined with respect to the rigorous rules of logic. Thus if mathematical knowledge is to be epistemically different from natural scientific one with no doubt the acting components that deliver the knowledge should be epistemically different as well. This naturally should include the self-correcting procedures in mathematics.

According to the principle of fallibility propositions have the property of being fallible. Their fallibility is examined through revisions. The general fallibility however is not epistemically specified and from an epistemic point of view it has to be since it is conducted by (revising) epistemic justifications and they always have an epistemic kind. Thus, the propositional property of being revisable is a potential property unless actual revision takes place. In this sense, revisions are the leading factor in both demonstrating the property as well as its materializing. The identification of a revision usually comes from its end result: when a proposition or a set of propositions are modified in some way it is natural to explore the actual reasons behind the modification. Thus, starting from the end proposition(s) and following through the reasons that led to it we can trace back the precise formulation of the initial proposition that got revised in the first place. Often, the end proposition is available together with the initial one so then the procedure of tracking the revision down consists just in the precise identification of the separate steps of the revision and their justification. In order to identify the epistemic kind of the revision we need to identify the epistemic kind of the justification of the propositions involved as well as the epistemic kind of the reasons to follow them. It is more or less clear that a requirement about completeness might be too strong to be achieved; for one might sometimes argue that another intermediate step or a proposition is necessary in order to pass to the next one or even in order to have a reason to subscribe to a step. Nevertheless, and especially in mathematics and sciences, the procedures are more or less discriminable and thus their epistemic kind is traceable as well. Revisions in physics have most of the time a clear-cut starting point, procedure and end result. Even if some additional philosophically interesting steps are perhaps acting tacitly, we can still inquire about the actual revision that took place. And actual revisions in fact are those which show us how scientific knowledge evolves. Thus, the worry about completeness and infinite regress are of more interest for hard core logic than for scientific epistemology.

Rules, Propositions and Epistemic Homogeneity

The obvious problem before the notion of a priori revisability is the epistemic route to the end epistemic kind of the revision. If a revision is identified and some of its steps, including, say, the revised proposition and the end modified proposition, are justified independently from experience but others are not, is it legitimate to regard the epistemic kind of the revision as an a priori one? The epistemic non-homogeneity of the components of the revision seems to threaten the epistemic claim of the notion. As before, the main source of help for settling such questions comes from the uncontroversial notion of empirical revisability. Being the complementary epistemic kind of revisability it seems justified accepting that whatever epistemic rules govern its homogeneity problems the same should work for a priori revisability as well. So, let us symmetrically translate the question: if in a revision the revised proposition and the modified proposition are justified empirically but some of the transitional steps seem to be justified independently from experience is it legitimate to accept that the epistemic kind of the revision is still an a posteriori one? What we have to consider here is the ability of the revision procedure to conduce epistemic kind reminding ourselves for the ability of a sound logical argument to conduce truth from premises to its conclusion. If we have to consider the *process* of the transition between the revised proposition to the transitional revision steps and then to the modified propositions then it seems that we cannot allow for a discontinuity to occur. Thus, even if though every proposition involved has its own justification that comes with its own epistemic kind, the epistemic kind of the overall procedure consists in building up an overall epistemic kind out of the constituent epistemic kinds. The presence of a different kind somewhere along the line would interrupt the epistemic passage to the overall epistemic nature of the revision. In this way it seems that an epistemically non-homogenous procedure could not transfer epistemic kind even if the initial and the end epistemic kinds of the steps of the revision are the same. In this sense, from a formally-epistemic point of view non-homogeneity seems to disallow for the identification of a unique overall epistemic kind of a revision.

Yet, this is not what we observe in actual cases of epistemic revisions. Sufficiently precise examination might find revision components that are independent from experience in probably even the best candidates for empirical revisions. To illustrate, all revisions are subject to logic and it is classical (or some modern) logic that governs the transition between the different steps. Thus if we take the revision of the Aristotelian thesis that the sun revolves around the earth ending at the Copernican thesis for the opposite we would see that the transition between the otherwise perfectly empirically justified steps is at least to a great extent a logical transition, that is, it is enabled by our trust in the principles of logic. Recently, Miscevic[27] stressed an influential objection against a similar kind of reasoning, commenting on Bonjour's influential argument against empirical justification.[28] He pointed that it is (as I understand it) a sort of category mistake to refer epistemically in the same way towards rules as if they are propositions. The distinction is certainly of crucial philosophical importance and thus of epistemological as well.

The difference between rules and propositions, however, is epistemically perhaps even more subtle. For to accept that the epistemic status of rules is different from the epistemic status of propositions is one thing and to accept that we cannot have or we do not have an epistemic attitude towards rules is a completely different thing. Because it seems that we do have an epistemic attitude towards rules, say, the rules of classical logic. We have identified them long ago in Antiquity and we follow them in present days. The fact and the procedure of their identification are instructive: for we have chosen *those* rules and not others and we have chosen so for good reasons and not arbitrarily. Thus, it seems that by having reasons to prefer some rules before others we are in fact justified in holding those rules in a sort of way very similar to the epistemic justification. For having reasons is what is at the heart of the epistemic justification and thus it seems quite legitimate to inquire about the epistemic nature of our reasons to subscribe to the rules. Transitivity of justification passes our reasons to hold the rules to reasons to hold the different specific steps of the rule that constitute them, along with the rule-like relations that connect them. If I have reasons to hold a rule then I have reasons to hold the constituents within the rule.

So far it does not seem that we differ in our general epistemic attitude towards rules comparing to our attitude towards mere propositions. Now, Miscevic might rightfully point again to the difference between rules and propositions and argue that even if we do have a legitimate epistemic attitude of the same kind towards both rules and propositions it is not of much help, for rules are *different* from propositions and thus we are working in two distinct domains. As a consequence, an argument I build within one domain need not hold for the other domain at all until the opposite is proven explicitly. True as it is, this would seem to understress the fact that every rule is rationally approachable only as a linguistic entity, that is, as a proposition, and only in its propositional formulation it is available for philosophical investigation. If we want to regard rules as not being propositions (besides being something else) this might endanger their position as linguistic entities and hence their being objects of linguistic and philosophical attitudes. For the rules are not merely terms, which they obviously employ, but they also connect those terms in some way and this is what usually linguistic propositions do. The fact that the rules *prescribe* distinctive steps to be followed conditionally is a property *complementary* to their property of being available as propositions and also, a property that is not at all easy to see as a linguistic property. And this is why, it seems to me, that it is legitimate to inquire about their epistemic status in their quality of being propositions besides the legitimacy to inquire about their epistemic status in their quality of being rules for following.

The difference between rules and propositions does not seem to be of a linguistic kind. All rules are rationally approachable, even given the fact of the discrimination between their separate steps, only as propositions. And it seems that in principle every well formed set of standard linguistic propositions could be elevated to the status of a rule by merely changing its modus. To take a simple example, the propositions "The hat of Julie is yellow" (Y) and "Marc drives fast" (F) seem like ordinary linguistic ones. If we try to transform them into a rule (the usefulness and the rational sense of such a rule is certainly far from clear but this is irrelevant to the fact that we *can* turn the propositions into a rule) we can say that the follower of the rule should do F if Y is present. Thus, to follow the resulting rule would be for Marc to drive crazy *whenever* Julie's hat is yellow. The steps of the rules are distinct enough: the first step is "the hat of Julie being yellow" (or not, as it were) and the second is "Marc driving fast". The following of the rule consists in the action of the follower who executes step 2 *given* step 1. The interesting fact about this is that nowhere during the transformation the initial set of propositions seems to lose its linguistic (or logical) nature. The property of being "a rule to follow" is thus an additional property of the well formed in natural language propositions Y and F. While taken as a rule the question of the epistemic justification of both Y and F seems not to be in focus. Yet, in closer inspection, following step 2 seems to depend on the availability of step 1, that is, on the fact that the hat of Julie is yellow. For if the hat is red Marc would not follow the rule; for there are no other steps available in the current rule-domain than Y and F and thus there would be no defined reason for Marc to execute a different step. The difference between the hat of Julie being yellow and the hat not being yellow is best captured in terms of truth and falsity. And truth and falsity are not to be had without epistemic justification. In this way the truth of step 1 seems like a necessary condition for the rule to be followed. In this scenario it is clear that we can inquire about the epistemic nature of the propositions that are constituent of a given rule and this reveals that question about epistemic justification is not at all ceasing to be directly relevant in the case of rules.

Therefore, even if a rule like *modus ponens* (in fact a complex proposition) actually governs the way a logical argument develops it is a rule that we subscribe to in its form as a proposition and thus we do seem to have epistemic route for our reasons to hold it. Even if *modus ponens* does not govern the transition from one step to another in its quality as a proposition it is nevertheless rationally and epistemically upheld through its *being a* proposition. And this is why it seems legitimate to attribute epistemic kinds to our reasons to hold it. If all of the above is correct then it seems that we can inquire about the epistemic kind of the reasons to pass from one step to another and as far as those steps are most of the time logical steps it seems natural to accept that at least some part of those reasons (for logical rules are part of our reasons to pass from step A to step B) are independent from experience, that is, that they are a priori. And in this sense even in cases of revision of purely empirical propositions we do seem to employ reasons to do so that are a priori and hence, it is difficult to see how an empirically conducted revision of empirically justified propositions could be homogenously empirical.

Yet, we do regard those revisions as empirical and not a priori. Therefore, the above assumption that for the sake of precision we should not regard non-homogenously conducted revision as having own distinct epistemic kind should be abandoned. For even if part of the reason and the means to perform a revision come independently from experience, this is not what constitutes the acting epistemic nature of the justification of the end result of the revision. And in case of empirical revision most of the time the modified proposition that is the end result of a revision is justified empirically. The revision is certainly most interesting as a procedure but at the end of the day for the scientific epistemologist it is the piece of knowledge that matters and this piece of knowledge is encoded in the end result of the revision, the modified proposition and not so much in the peripheral windings of the route that leads to it. The transitional steps are certainly also of utmost epistemic and logical interest but the justification to subscribe to them as a whole together with the propositions involved certainly does have a clearly identifiable epistemic kind that dominates.

The upshot of all this is that the epistemic route that does most of the job is the best candidate for the epistemic kind of the end result of the revision and in empirical revisions this is naturally experience (or scientific experience). Therefore, I take the homogeneity requirement to be too strong to subscribe to and as a consequence, I take it that we can in fact identify a predominant epistemic kind of a revision even of there are epistemic fluctuations in the actual procedure. This seems to be in harmony with what actually goes on in epistemology and science: empirical revisions are de facto regarded as empirical even if under scrutiny logical and mathematical dependencies are to be found almost all over the place. I believe that even the empiricist would agree that epistemic homogeneity is not a *conditio sine qua non* for the identification of a pure epistemic kind of be it justification or revision. For if she disagrees then she would face a dilemma, that would be pretty complicated to resolve: either homogeneity requirement is kept and thus it is quite difficult for the apriorist to establish purely a priori revisions but also, it is quite difficult for the empiricist to establish purely empirical revisions, or, the homogeneity requirement is not kept and empirical revisabilities are much easier to pronounce but so are a priori revisions.

Thus, the above considerations taken into account, it is sufficient to identify a revision as predominantly a priori (or empirical, as it were) in order to be legitimate to regard it as an a priori revision. Epistemically, this facilitates the task before the apriorist for now she can more or less safely neglect epistemic fluctuations as far as she manages to establish a predominant epistemic kind for a revision.

Unique Epistemic Revisions?

Perhaps the most interesting epistemic question concerning a priori revisability is about its relationship with empirical revisability. What about cases which are supposedly a priori revisable, are they *uniquely a priori revisable* or they are empirically revisable as well? Much of the epistemic strength of a priori revisability seems to stem from this question. For if an a priori revisable proposition turns out to be also empirically revisable this undoubtedly would diminish its overall independence from experience. Therefore, the apriorist should attempt to clarify the relationship between a priori and empirical revisabilities as precisely as possible. Again, it would be most instructive to see what the situation with the empirical revisability is.

Is an empirically justified proposition that is empirically revisable not entirely dependent on experience, because it is also a priori revisable? Technically, the question has to distinguish between potential revisability and actual revision. For if a proposition is only potentially revisable this does not seem to threaten its actual overall epistemic character. On the other hand, if a proposition has actually been revised in some sort of epistemic way this does seem to have a direct bearing on the actual overall epistemic kind of the proposition. The question about the actual epistemic nature of a proposition might be related to but is different from the question about the possible ways for its epistemic revision. In order to spell out the latter we need to delineate the general epistemic conditions for revisability and then to examine any given proposition with respect to them. In the case of empirical revisability the proposition is capable of being first revisable and second, empirically revisable. Whereas its property of being generally fallible is delivered directly from the principle of fallibility its property of being empirically revisable is not. For empirical revisability seems to involve more than mere possibility the truth value of the proposition to turn out to be different. There has to be a connection between the subject matter of the revisable proposition and the subject matter of the revising proposition (or propositions). Thus if we take again the revisable proposition "The Sun revolves around the Earth" and look at its subject matter it is clear that there has to be some sort of relation between the things the proposition is about, that is, sun, earth and revolving. In empirically justified propositions this is the causally regulated experience of the cognitive agent with the alleged referents of the singular terms of the proposition. Now, if we take the revising proposition (actually as history shows the revision was immensely more complicated and involved a whole lot more propositions, but for the sake of simplicity and making a clear point I would take the simplest case arguing as if the revision is performed by a single proposition that challenges the revisable one) "The Earth revolves around the Sun" and if we look at its subject matter we would see that there is not much difference between the two. For again there seems to be present some sort of relation between the terms of the proposition and its alleged referents. But most importantly, the connection is of the *same kind*. For the kind of the supporting justification behind the revisable proposition is empirical and this is clearly due to the nature of the subject matter of the proposition: we need to have causal experience with the Sun and the Earth on order to be justified in holding the proposition (provided that the experience in question is, of course, consistent with and supports what the proposition is saying). And if we have the same subject matter, as we actually do in the case with the revising proposition, it seems that we have to have the same relation to it, for otherwise the proposition would have been justified in a different way.

For a clear-cut case like the present one this is not what happens and the revising proposition is empirically justified in the same sort of way like the revised one. Obviously, the experience in both cases is different, for in the first case it supports one proposition and in the second its rival. But the experience is different as *content* and not as a *kind*; it is the same epistemic route. The obvious reason for that is that there is a working epistemology which governs our knowledge about propositions like this and this is empiricism. The reason why empiricism *is* the working epistemology here is at least twofold. First, this is how actual human knowledge works when presented with questions about entities which are accessible only (or mostly) through empirical observation and secondly, and more importantly, because there is no other conceived of epistemology that could deliver the knowledge that we actually do seem to have. The very nature of planets and stars, as we are aware of it, is such that it seems it cannot be known unless we are observing directly or indirectly the planets and stars. Therefore, it seems that *the working epistemology is in fact to a great extent determined by the subject matter of the propositions*. Again, this is the pathos behind Bob Hale's recent stress of the importance of the nature of the entities that are subject of our theories.[29] Further, it is surely imaginable that several epistemologies are possible to deliver knowledge encoded in some P. Yet in cases like the above it seems that is only that doing the job. This, among other things, is as good a commercial for the epistemology as it gets. And the mere imaginability should do much better than just allow for an abstract possibility; for it is not clear how abstract possibility could reveal a rival epistemology as actually capable of accessing the same type of propositions. Epistemically viable imaginability should be regulated by non-abstract conditions of possibility and this is much more difficult to obtain. The upshot of this is that the subject matter of propositions does have some nature which determines the spectrum of possible cognitive routes to it. And clearly the epistemic justification of the proposition, that has the merit of delivering knowledge, should have followed one of those routes. Unfortunately, there seems not to be many possible epistemic routes to the traditional subject matter of humane knowledge. Thus, epistemic justifications are not in a position to be too picky.

Illustrations from empirical revisability again demonstrate their explanatory potential. For if we start with propositions that we are clearly most willing to accept as delivering knowledge, fallible as they are, we can trace back the working epistemology that actually delivers the knowledge. Even more, by tracing the epistemology we can identify those aspects of the nature of the entities to be known that determine the types of cognitive access an agent could *actually* have. Thus we can end up with what seems like a pretty good picture of the epistemic mechanism that delivers the knowledge we started from. And the most important component in this mechanism is the very nature of the things to be known for it is this that determines the ways we could follow in order to get knowledge. In paradigmatic empirical cases this is most clear. Yet, the same mechanism seems to deliver cases of knowledge that are pretty different in epistemic kind from paradigmatic empirical cases. And for two reasons. First, obviously the actual route of delivering knowledge for given proposition (like mathematical and logical) is not the same as paradigmatic empirical route and second, again and more importantly, the very nature of the entities to be known is such that empirical access to it is inconceivable. As we have seen above with illustrations of mathematical propositions we do seem to have knowledge that sum of angles in a triangle for all triangles (drawn or imagined in Euclidean space) is always 180 degrees and yet the way we seem to know this does not reveal any observation or measurement be it of triangles in actual diagrams or in mental imagination. Also, as far as the proposition "The sum of the interior angles of all (Euclidean) triangles is always 180 degrees" *is about triangles and angles* and it is really difficult to see how it is not about them, and as far as the true referents of the very notions of "triangle" and "angle" are not part of the physical world, which is the usual domain of empirical accessibility, it is clear that the nature of the entities, referred to by the terms in this and similar propositions are of *a different kind* when compared with planets and stars. Thus three things are clear: first, that we do have knowledge delivered through (true) mathematical propositions, second, that we do seem to arrive at this knowledge in a way different from standard empirical way and third, the nature of the things the propositions are about is such that does not seem to allow empirical access to them in the first place. In this way we have what is more or less clear an alternative epistemic mechanism to the above illustrated empirical one.

What is the relation between these two mechanisms and the epistemic revisability conditions of propositions? For starters, if the nature of propositional referents determines to at least some extent the possible ways of cognitive access to them, then it is clear that propositions that concern entities cognitively accessible in a way 1 would have different supporting and revising conditions than propositions that concern entities cognitively accessible in a way 2. In other words, if we have a proposition about empirically inaccessible entity and in quantum physics it is not very difficult to get one, but still prefer that we do have some certain knowledge about them, we would have to conclude that the epistemic way we reach this knowledge is different from empirical. Whether the nature of the entity in question is *uniquely* empirically inaccessible is a million dollar question. In physics cases it is clear that some entities that were empirically inaccessible in, say, year 1920 became empirically accessible in year 1935 or later due to the rapid development of technologies (the elementary particle known now as "positron" was not empirically accessible at the time when Paul Dirac actually predicted its existence;[30] it nevertheless became empirically accessible some years later when the first positrons were actually observationally detected, although in what we would accept to be an indirect way). In this sense, it seems reasonable to accept that the entity was not uniquely empirically inaccessible even in year 1920. On the other hand if we agree that mathematical propositions deliver knowledge and are about abstract entities it seems pretty difficult to conceive of a situation in which those entities would become empirically accessible. In this sense abstract entities are much better candidates for uniquely a priori accessible than the standard physics elementary particles. Yet, even in empirical science it seems that some physically existing entities would very likely never become empirically accessible to us no matter how good technological development we enjoy. For according to our best theories the nature precludes of the possibility to accessing some of its fundamental constituents. Thus, to leave Heisenberg's uncertainty considerations aside, it seems that even if string theories happen to get it more or less right about strings being the ultimate atoms of the physical universe we would most probably be never able to probe that deep. Not because it is too expensive to build the collider that would show the strings to us but because the laws of nature seem to disallow for such probing in principle. Yet, we would still prefer to believe that we nevertheless could have some knowledge about those entities and not just in order to avoid skepticism. Most physicists would prefer to accept that in case we got the laws of nature more or less right we can still get valuable knowledge even about empirically inaccessible entities of the ultimate kind. Besides all, it is not far from the actual way scientific knowledge evolves. A brilliant illustration would be James Robert Brown's epistemic analysis of the Eistein-Podolski-Rosen (EPR) thought experiment in quantum mechanics.[31] The way I take it Brown manages to demonstrate in a most clear philosophical way that we not only could but we also do have knowledge about entities without being in any sort of epistemically viable causal contact with them. And as far as pretty much any modern available account of empirical knowledge rests in standard notion of causality this amounts to actually delivered knowledge that is independent from experience, that is, a priori. Last illustration could be notorious propositions about non-existing objects. If we agree that the proposition "Unicorns have one horn" is *about* unicorns and not about, say, rabbits, and if we agree that it does differ quite interestingly from the proposition "Unicorns have five horns" then it seem that we have to agree that for a physically non-existent entity like the unicorn this proposition does deliver some epistemically interesting knowledge. And if it does clearly it could not have come through any sort of experience for there are no unicorns and consequently, there is nothing for the cognitive agent to be causally in a contact with. In this sense, if we prefer to accept that we do indeed have knowledge about cases like this and also for cases where propositions concern future events we better conceive of alternatives to causal empiricism if we do not want to fall into skepticism.

In this sense it is the very nature of the subject matter of propositions that determines epistemic conditions of justification and revisability. It is beyond our present purposes to suggest an exhaustive classification of actual epistemic schemes but it is perhaps relevant to provide a set of rules to judge when presented with actual cases. First, the nature of the entities the given proposition is about has to be determined and the question should be answered whether those particular kinds of entities are opened for empirical access. Second, the proposition should be examined with respect to the probability of delivering knowledge: if we are strongly unwilling to deny that a proposition is actually delivering knowledge then it seems that we must comply with the cognitive routes of access to its entities we are allowed by them. Third, every effort should be made to discriminate between mixed and clear-cut cases. Prima facie, however, it seems conceivable that both mixed scenarios have a great degree of plausibility. It is equally plausible to accept that empirically accessible entities might be a priori accessible as well (Brown's illustration of *EPR* comes in pretty handy here; for properties of the remote (causally inaccessed) particle were determined thought experimentally (although fallibly) before technologies actually permit empirical observation. In the mid-eighties, some 50 years later, the French physicist Alain Aspect actually conducted the thought experiment as a real physical experiment and some empiricist might see in this empirical accessibility of the particle)[32] as to accept that some a priori accessible entities might turn empirically accessible as well. Thus the question about uniqueness of epistemic revisability seems best answered *in concreto* and not in general.

Are we to regard then actually a priori revisable propositions as not so strongly independent from experience merely because the entities they are about might turn empirically accessible in, say, 20 years? It does not seem so. For when we are presented with a proposition justified empirically about those same entities we would have a case of empirical justification and when we revise our empirically justified proposition in an empirical way we do have empirical revisability. In this case we would not argue that the proposition is not entirely dependent on experience both through its justification and its revision merely because it is possible also to be revised (or justified) independently from experience. Dependence on experience seems more actual than potential: in order for the potential one to approach the significance of the actual one it has to be demonstrated as being on an equal footing with it. And it is not immediately obvious how this could be for one of the reasons for a dependence on potential and not actual is because the knowledge did not follow this way. The reasons might certainly be arbitrary but still, actual epistemic justification and potential ones seem to differ in strength. In order for the epistemic kind of a proposition to be demonstrated its actual dependence on the independence from experience should be demonstrated. Potential one does not seem to do. When potential cases turn actual the epistemic kind of the proposition is reestablished and in this sense it might be regarded as a new one. The empiricist would not suffer from alleged independent from experience justification if it is potential: for no cognitive agent actually conveyed such justification. Thus the symmetric case should be granted as well.

Epistemology should not strive for the ultimate modal picture where every possible scenario is clearly examined epistemically. This seems too complicated an equation to be compiled let alone resolved and the conditions of its possibility are far from obvious. Epistemology could instead come up with criteria for actual independence from experience and judge cases according to them. The obvious candidates are the following:

(1) When a proposition is about entities which nature is such that there is too little or no probability to become empirically accessible (as mathematical and logical cases, some fundamental cases from physics perhaps as well) and yet they seem to deliver knowledge in a perfectly fallible way it seems plausible to accept that the conditions of their revisability (as far there are some) would be uniquely a priori

(2) When a proposition is about cognitively mixed entities which are both empirically and a priori accessible it seems to be useful to look at actual cases of justification and revision: for a proposition is still a priori revised and revisable even if it is conceivable to be empirically revised in few years.

(3) When a proposition is about entities which nature is such that they are predominantly empirically accessible it is pretty safe to expect that the conditions of its revisability would be also empirical. For the very subject matter of the propositions needs to be addressed in the same sort of access way. Which is different from denying that at least in principle, it might turn out to be accessed in a different way.

The epistemic upshot of this is perhaps the following: knowledge, scientific included, seems to be able to evolve through empirically independent moves. Fallibly, by all means. But not differently fallibly than empirical moves themselves. The value of this aspect of epistemic dynamics for science and mathematics could hardly be exaggerated: both empiricist and rationalist like to have the progress of knowledge and both epistemic kinds are so intimately intertwined within themselves that sometimes it seems irrelevant to inquire about their epistemic nature. Even where they disagree the total body of results of the scientific enterprise seems to continue its evolution. Nevertheless, epistemic dynamics does seem to have a rationalistic aspect and this is of certain interest for the philosophically minded.

Notes

---

[1] Reichenbach, Hans [1965] *Theory of Relativity and A priori Knowlegde*, Los Angeles, University of California Press.

[2] Carnap, R. [1936] "Testability and Meaning" in Philosophy of Science, III (1936) and IV (1937); Carnap, R. [1947] *Meaning and Necessity*: a Study in Semantics and Modal Logic, Chicago : University of Chicago Press; Carnap, R. [1928] *The logical Structure of the world*. Berkeley and Los Angeles: University of California Press.

[3] Kant, Immanuel [1781/87] *Critique of Pure Reason*. Trans. Norman Kemp Smith as *Immanuel Kant's Critique of Pure Reason*. London: Macmillan Co. Ltd., 1963.

[4] Quine, W.V.O. [1951] "Two Dogmas of empiricism" in *The Philosophical Review* 60 (1951): 20-43.

[5] Putnam, H. [1979] "What is mathematical Truth" in *Philosophical Papers*, Vol. I, CUP, pp. 60 - 78; Putnam, H. [1983] "Two Dogmas Revisited" in Philosophical Papers, Vol III, CUP, pp. 87 – 97.

[6] Kitcher, Philip [1980] "A Priori Knowledge," *Philosophical Review,* 89: 3-23.

[7] Pierce, Charles: Collected Papers of Charles Sanders Peirce, 8 vols. Edited by Charles Hartshorne, Paul Weiss, and Arthur Burks (Harvard University Press, Cambridge, Massachusetts, 1931-1958).

[8] Kripke, Saul [1972] "Naming and Necessity", In *Semantics of Natural Language*, edited by D. Davidson and G. Harman. Dordrecht; Boston: Reidel.

[9] Casullo, Albert [1988] "Revisability, Reliabilism and A priori knowledge" in *Philosophy and Phenomenological Research*, 49, pp. 187 – 213 and Casullo, Albert [2003] *A priori Justification*, Oxford University Press, NY.
[10] Miščević, Nenad [2001] "Apriority and Conceptual Kinematics" in *Croatian Journal of Philosophy* (1 (1)/2001), pp. 21-48.
[11] Kripke's *Naming and Necessity* is a prominent illustration.
[12] Mill, John Stuart [1843] *A system of Logic*, London, Longmans.
[13] Frege, Gottlob [1885] *Foundations of Arithmetic*, Oxford, Blackwell.
[14] Kitcher, Philip [1983] *The Nature of Mathematical Knowledge*, OUP.
[15] It is important to present the pebble counting case with sufficient precision for it usually provides grounds for epistemic confusion. Thus when an empiricist like Mill says that I am justified (to use the modern formulation) in holding a proposition by counting pebbles we must inquire about *which* proposition precisely he is to be justified in holding. For the proposition "Two pebbles plus three pebbles equals five pebbles" (P) is quite different from the proposition "Two plus three equals five" (Q). Even if we agree that he is empirically justified to hold P it is far from clear how by virtue of the same very experience he is at the same time justified in holding Q as well. Also, it is far from uncontroversial that it is *only* the given experience that manages to justify him holding even P. For obviously, we have a coordination between a theorem of pure mathematics (2+3=5) with a set of selected physical objects and the *assumption* that the truth of the theorem somehow transfers to the given selection. The experience of counting the five pebbles is obviously insufficient to deliver this coordination; and quite naturally so, for the problem of the application of mathematics is one of the notoriously difficult ones to solve and not just from the time of Hilbert. At the end of the day it is clear that the empiricist could not be justified in holding Q and has serious problems to count as being justified in holding even P, irrelevant as it happens to be for epistemology of pure mathematics.
[16] For reduction of geometry to arithmetics see Hilbert, David, [1918] "Prinzipien der Mathematik", Lecture notes by Paul Bernays. Winter-Semester 1917-18. Unpublished typescript. Bibliothek, Mathematisches Institut, Universität Göttingen and Hilbert, David, [1922] "Grundlagen der Mathematik", Vorlesung, Winter-Semester 1921-22. Lecture notes by Paul Bernays. Unpublished typescript. Bibliothek, Mathematisches Institut, Universität Göttingen.
[17] Hale, Bob [2007] in the introduction of "The Problem of Mathematical Objects", p. 1, forthcoming; (delivered as a talk at the Central European University in November 2007).
[18] Casullo, Albert [2003] *A priori Justification*, OUP.
[19] Kitcher, Philip [2001] "A priori Knowledge revisited" in Bogghosian and Peackock (eds.) *New Essays on the A priori*, OUP.
[20] The clearest exposition of the two senses is perhaps due to Casullo in his Casullo, Albert [2003] *A priori Justification*, OUP.
[21] Casullo, Albert [1988] "Revisability, Reliabilism and A priori knowledge" in Philosophy and Phenomenological Research, 49, pp. 187 - 213.

[22] Edidin, Aron [1984] "A priori Knowledge for fallibilists" in *Philosophical Studies*, 46, pp.189-97.
[23] Casullo, Albert [1988] "Revisability, Reliabilism and A priori knowledge" in Philosophy and Phenomenological Research, 49, pp. 187 – 213 and Casullo, Albert [2003] *A priori Justification*, OUP.
[24] Edidin, Aron [1984] "A priori Knowledge for fallibilists" in *Philosophical Studies*, 46, pp.189-97.
[25] Casullo, Albert [1988] "Revisability, Reliabilism and A priori knowledge" in *Philosophy and Phenomenological Research*, 49, p. 190. Most recently he discusses the notion at greater length in Casullo, Albert [2003] *A priori Justification*, OUP.
[26] Thurow, J. [2006] "Experientially Defeasible A priori justification' in *The Philosophical Quarterly*, Vol. 56, No. 225, pp. 596 – 602.
[27] Miščević, Nenad [1998] "The Rationalist and the Tortoise" in *Philosophical Studies*, Springer, Issue Volume 92, Numbers 1-2 / October, 1998 Pages 175-179.
[28] Bonjour, L. [1998] *In Defense of Pure Reason*, Cambridge University Press, UK.
[29] Hale, Bob [2007] in the introduction of "The Problem of Mathematical Objects", p. 1, forthcoming.
[30] Paul Dirac predicted the existence of the positron in 1928 as a consequence of the Dirac equation.
[31] Brown, James Robert [1992] "EPR as a priori science" in *The Return of the A priori*, Supplement to the Canadian Journal of Philosophy
[32] Ibidem. For discussion of Aspect experiment see A. Aspect, P. Grangier, and G. Roger [1982] "Experimental Realization of Einstein-Podolsky-Rosen-Bohm Gedankenexperiment: A New Violation of Bell's Inequalities" in *Physical Review Letters*, Vol. 49, Iss. 2, pp.91-94 (1982) doi:10.1103/PhysRevLett. 49.91

CHAPTER V

THE REVISION OF THE FIFTH POSTULATE OF EUCLIDEAN GEOMETRY AS AN A PRIORI REVISION IN GEOMETRY

The chapter presents an analysis of the revision of the fifth postulate of Euclidean geometry as an illustration of an a priori justified revision. The first section extracts the idea of a priori revisability as a natural epistemic counterpart of the idea of empirical revisability. Section two gives the history of the attempts to prove the postulate and analyzes the changes that affect its truth and meaning. Section three clarifies the exact subject of the revision balancing between the revision of the individual proposition and the revision of the geometrical system. It also delineates the different stages of the revision. Section four discusses the revision in the light of the distinction pure-applied geometry. Section five presents the epistemic analysis of the revision, gives the main argument for its a priori nature and responds to some relevant objections.

History of science is rife with examples of empirical revisions of scientific beliefs. Virtually every specific science uses observation and measurement to test its hypotheses and to revise accepted propositions; modern genetics uses sophisticated microscopes in order to get to the real properties of the human genome and modern astronomy uses state of the art telescopes in order to probe into the far reaching regions of the universe. Technically speaking even the gathering of entirely new knowledge through observation might count as a revision: of the older believes that (eventually) addressed the issue. To illustrate, until fairly recently astronomers believed that there are 9 planets in the Solar system. As it turned out, the very concept of a planet had to be revised and hence all the propositions it figured in as well; and all this due to the new findings about the existence of unsuspected cosmic objects whose observed properties like size, mass and location in space interfered with the definitions of contemporary astronomy. Thus, it seems to be beyond question that the role of human and scientific experience is indispensable for contemporary science. The standard empirical scientific revisions have roughly the following (presented in a simplified manner) structure:

P – proposition (or set of propositions, most often within a theory) to be revised
Q – proposition (or set of propositions) that address P's content in a way, different than P
Process of evaluation of P and Q (within a theory),

P* - modified P (or R, P is substituted with a new proposition)

The revision has an epistemic aspect. Thus both the justifications behind P and Q and the justification for the very revision have epistemic kinds. In a standard scientific example P is maintained on empirical grounds and has a justification J_e; in the general case Q is also held on empirical grounds and has justification J_e. Since the source of the reasons to accept a revision is experience, the revisionary reasons (so far unidentified precisely) are also of empirical character. In this way we receive the following epistemically reformulated structure of a typical revision:

P with Je – proposition (or set of propositions, most often within a theory) to be revised
Q with Je – proposition (or set of propositions) that address the P's content in a way, different than P
Process of evaluation of P and Q (within a theory), governed by empirical considerations (Je)

P* with Je - modified P (or R, P is substituted with a new proposition)

Clearly, the grounds to call revisions of this kind "empirical" are in the epistemic kinds of the participants in the revision. The extracted epistemic structure of the empirical revision is

Je for P
Je for Q
Je for the revision

Je for P* (R)

This structure, however, does not do justice to the available epistemic kinds for there is one more epistemic kind, the complementary to the empirical a priori kind. In this sense, a purely formal epistemic structure of a revision that addresses the epistemic kinds as variables and not as filled in values as in the case above should be given as

J_k for P
J_k for Q
J_k for the revision

J_k for P* (R)

Where "k" stands for "an epistemic kind". In this way, if we follow the structure of the epistemically uncontroversial empirical revision, and if we also accept the mere availability of the a priori epistemic kind we end up having a general formal structure of an epistemic (scientific, in this case) revision. Thus, it is obvious that other epistemic options besides the empirical one are conceivable: the homogenous counterpart would be a purely a priori revision of the kind

Ja for P
Ja for Q
Ja for the revision

Ja for P* (R)

where "a" stands for the "a priori" epistemic kind of the justifications. Epistemically non-homogenous combinations are certainly conceivable as well:

<table>
<tr>
<td>Ja for P
Je for Q
Ja for the revision

J? for P* (R)</td>
<td>Ja for P
Je for Q
Je for the revision

J? for P* (R)</td>
<td>Je for P
Ja for Q
Ja for the revision

J? for P* (R)</td>
<td>Je for P
Je for Q
Ja for the revision

J? for P* (R)</td>
<td>Je for P
Ja for Q
Je for the revision

J? for P* (R)</td>
</tr>
</table>

Table 5-1

Prima facie, it is not clear what the epistemic kind of the non-homogenous combinations should be: empirical, because of the participation of an empirical justification, a priori, because of the participation of an a priori justification, empirical, because the empirical justification overtakes epistemically the a priori one, a priori, because the a priori justification overtakes epistemically the empirical one, etc. What is more significant, however, is that we are presented with an intuitive model for epistemic classification of cases of revision that is based on the epistemic kinds of the participants. Thus, in the empirical homogenous case the empirical kinds of the justifications are sufficient to classify the revision as empirical one; hardly any natural scientists would object to this kind of reasoning. As a consequence, however, and at least logically, we can conceive of a priori homogenous cases of revision, where the epistemic kinds of the participants are a priori. For the time being we might bracket the non-homogenous cases and discuss the clear-cut ones.

In science the mere claim for logical conceivability is not of much help, for science is interested in how things actually are and not in just how we can conceive them to be. In this sense, an epistemic evaluation of a scientific revision is not particularly useful from a scientific point of view, if it does not show to be an actual one, that is, to occur in the proper scientific practice and to be valid for scientific propositions, hypotheses and theories. After all, the empirical epistemic model of a scientific revision is extracted from the scientific practice; to a great extent this determines its influence within science. Therefore, if a rival epistemic model, as the a priori model of scientific revisions, has to be on a par with the empirical one, it must demonstrate its vitality in *actual* scientific cases. In the present and in the following chapter I will present and examine two case studies that would attempt to show that a priori epistemic revisions are not merely conceivable but also actual in history of science and mathematics. The first case would be an influential revision in mathematics, the revision of the Fifth postulate of Euclidean geometry; the second case would be an influential revision in natural science performed through a famous thought experiment in physics, Einstein's *Train Thought Experiment*.

History of the Revision of the Fifth Postulate

There is perhaps no other proposition in the whole of the history of mathematics that has become so famous and in a way, so notorious, as the fifth postulate of the Euclidean geometry. In the *Elements*, Euclid introduced 23 definitions (̓Όροι), 5 common notions (Κοιναὶ ̓έννοιαι) usually called "axioms" and 5 "requests" (Αἰτήματα) usually called "postulates" in English.[1] These were meant to provide all the necessary information and rules for the whole body of the geometry to be derived following deductive reasoning. In this sense, those propositions were of a different kind from the theorems: they were accepted and not proved whereas the theorems were proved deductively from them; the axioms (for the sake of simplicity further I will follow Reichenbach and others who refer to all the definitions, axioms and postulates under the common name "axioms") made the proofs possible in the first place. Clearly, they were accepted by virtue of some valuable property of theirs which enabled them to do the job and which, in doing so, distinguished them from all the other propositions of geometry. In the case of the axioms proper this property is often taken to be the immediate obviousness of their statement.[2] In the case of the definitions one might argue that it is the fundamental meaning of the building concepts that figured in the geometrical propositions.[3] In the case of the postulates it is perhaps the sort of the visual self-evidence of their statements that provides the property of a value. Above all, however, was the following common property of the axioms: that none of them could be deduced from the others and that all together they were sufficient to enable the deduction of all geometrical propositions in such a way that no proposition proved from the axioms (or from other theorems for that matter) would ever contradict any other.[4] There was one axiom, however, that looked different from the others. This was the famed fifth postulate, which looked so much more complex than the other postulates that geometers started to doubt whether really it cannot be deduced from them:

The Five Postulates of Euclidean Geometry

First Postulate: To draw a line from any point to any point.

Second Postulate: To produce a finite straight line continuously in a straight line.

Third Postulate: To describe a circle with any center and distance.

Fourth Postulate: That all right angles are equal to one another.

Fifth Postulate: That, if a straight line falling on two straight lines makes the interior angles on the same side less than two right angles, the two straight lines, if produced indefinitely, meet on that side of which are the angles less than the two right angles.

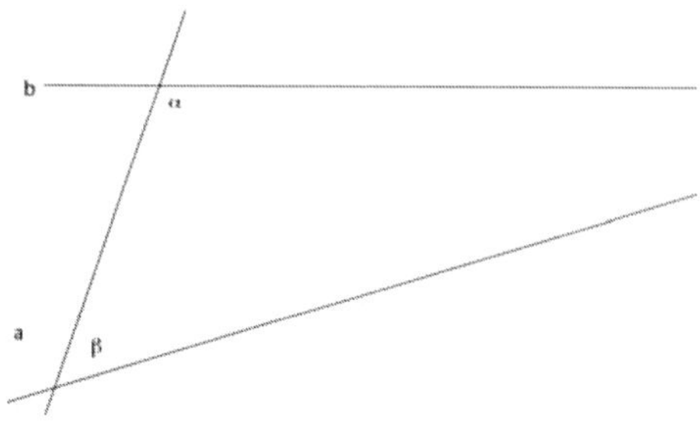

Figure 5-1

The attempts to prove the postulate from the other four are probably as old as the postulate itself. It is perhaps not very far from the truth to suppose that Euclid, the very creator of the *Elements*, was the first to feel an apparent uneasiness about the postulate. An illustration in support of this conjecture is the fact that he worked out many of the first theorems in the *Elements* with no appeal to the postulate. We know as early as 5th century B.C. about actual historical attempts to prove the fifth postulate. Proclus specifically mentions that Ptolemy, for example, has come up with one which turned out unsuccessful.[5] The subsequent history of geometry is full with a great number of recorded and unrecorded attempts to prove the postulate. Some of the proofs were even believed to be correct and sometimes it took quite some time to expose them. The problem became notorious in that turned into obsession for many a great mathematician: as an illustration, the mathematician Legendre devoted 40 years to its solving. All the proofs until the beginning of the 18th century failed, typically for being circular: in their structure they were incorporating a premise that concerned a property which was taken as geometrically evident, and later it inevitably became clear that this premise was actually logically equivalent to the proposition to be proved, the fifth postulate. During the late 17th and the early 18th century, however, a different kind of proof was attempted. The Italian Giovanni Saccheri attempted the technique of *reductio ad absurdum*: he assumed that the postulate is false and tried to reach a contradiction. In his work from 1733 *Euclides ab omni naevo vindicatus sive conatus geometricus quo stabiliuntur prima ipsa universae geometria principia* [6] Saccheri examined in depth the logical consequences of rejecting the fifth postulate. He proved many theorems in the resulting new axiomatic settings, but ironically, he was not aware of the fact that he has practically formulated (or discovered for that matter) an entirely new geometrical system. Eventually, while working in the system of hyperbolic geometry, he decided that he reached a contradiction whereas in fact there was no logical contradiction. After Saccheri the

Swiss-German Lambert was another famous mathematician who attempted a *reductio* at the fifth postulate.[7] Interestingly, he never reached a contradiction too but in the literature the historians of mathematics again do not seem to accept this as a factual discovery of the non-Euclidean geometry. The real discovery of the non-Euclidean geometries is associated with the names of Carl Gauss,[8] János Bolyaj,[9] Nikolai Lobachevsky[10] and Bernhard Riemann[11]. The German mathematical genius Carl Gauss followed the *reductio* model of Saccheri and Lambert and investigated the properties of a geometrical system where the fifth postulate was rejected and substituted with the assumption that there is more than one line through a point, not on a given line, that are parallel to it. In fact, this assumption is one of the several possible ways to reject a geometrical proposition which is equivalent to the fifth postulate, the so called Playfair's Axiom

> Given a line and point not on the line, it is possible to draw exactly one line through the given point parallel to the line.

Although baptized after John Playfair,[12] who suggested the axiom in the late 18th century, this proposition is known as equivalent to the fifth postulate from the antiquity and historically it has often been referred to as the postulate itself. It is also worth mentioning that Gauss did not publish his work and history tells us that perhaps this was due to the reigning authority of Kant at the time who regarded Euclidean geometry as a paradigmatic example of certain and true knowledge. Be that as it may, the young Hungarian mathematician János Bolyai was the first to accept that a new geometry, different from the Euclidean one, is actually possible. He removed the fifth postulate from the axiomatic system of geometry and substituted it with (as it turned out later one of) its negation(s). The negation which Bolyai used was the one that leads to a hyperbolic version of the non-Euclidean geometries; in this system there is more than one line parallel to a given line that pass through a given point that is not on the line and the sum of the interior angles of the triangles is less than 180 degrees. The same system was reached virtually simultaneously by the Russian mathematician Nikolai Lobachevsky (actually the geometry is dubbed "Lobachevskian geometry" for Lobachevski was the first who published although, historically, János Bolyai was the first who formulated it). Interestingly, Lobachevski followed completely independently the same logical approach like Saccheri, Lambert, Gauss and Bolyai: he rejected the fifth postulate and plugged in on its place one of its possible negations (that there exist two lines parallel to a given line through a given point not on the line). Some twenty years later, one of the doctoral students of Gauss, Bernhard Riemann, formulated a system of potentially infinitely many non-Euclidean geometries, thus introducing an entirely new geometrical field, based on the concepts (in their modern formulation) of curvature, manifold and Riemannian metric. The more than two millennia long tradition of attempts to expose the fifth postulate as a hidden theorem ended in the surprising creation of a system of rival geometrical systems. The new systems, however, did not posses a proof for their consistency until the works of Eugenio Beltrami and Felix Klein. Beltrami came up with a geometrical model of a two-dimensional hyperbolic geometry within the very system of Euclidean geometry. In this way he provided historically the first proof of relative consistency: the problem for the consistency of the hyperbolic geometry has been reduced to the problem of the consistency of the Euclidean geometry. A few years later Felix Klein provided the same type of relative consistency proof for the system of elliptical geometry. As a result, the quest for the proof of the fifth postulate has finished revising radically the human geometrical understanding.

The Exact Subject of the Revision

The long and complex history of the attempts to prove the fifth postulate might undoubtedly be assessed from different philosophical perspectives. One aspect, however, is of significant importance: that some substantial and apparently complex change actually occurred with respect to the Euclidean geometry and its fifth postulate in particular along the history of these attempts. It is this change that is targeted by the suggested in this chapter revision. An analysis that claims that a revision has been made has as its primary task to identify correctly the proposition or the set of propositions that have been revised. Also, it has to identify the very revision, that is, the change that concerns the proposition (or the set of propositions), its meaning, its truth value or its position and function within the geometrical system in which it is formulated. The fifth postulate is one of the fundamental principles of the Euclidean geometry, an unproved but accepted principle that concerns relations between lines, planes, angles and points. The postulate affirms unequivocally something about those primitive geometrical notions and does so in a way that appeals to a notion of an indefinite or infinite extension (production) of a line. The postulate receives its philosophically interesting meaning only within the system of the Euclidean geometry, that is, if we take it in isolation it might not be regarded as sufficiently well formulated because, for example, there would be no available definitions of the primitive geometrical terms employed (like line, angle, etc.). In this sense, the geometrical meaning of the postulate is given only within a broader system of propositions; in the historically discussed case this is the system of Euclid's *Elements*. Hence, any change of the meaning of the postulate would automatically lead to a change of the overall meaning of the geometrical system. The table below tries to identify the different stages of the change that concerns the postulate and also tries to specify the characteristics of the change with respect to the notions of meaning, truth-value and position within a (geometrical) system:

Stages of Change of the Fifth Postulate

STAGE	CHANGE
(1) Original formulation of the postulate within the *Elements*	Original meaning of the postulate Original truth-value of the postulate (true in the Euclidean system of geometry) Original function in the Euclidean system (a postulate)
(2) Undermined trust in the postulate as a type of geometrical proposition: postulate or theorem	Unsettled fluctuations in the attributed meaning of the postulate No change in its truth value (innocent until the opposite is proved) No change in the position of the postulate within the Euclidean system (innocent until the opposite is proved)
(3) Circular attempts to prove that the postulate is actually a theorem (from Ptolemy to Legendre)	*Unsuccessful*: Unsettled fluctuations in the attributed meaning of the postulate No change in its truth value (innocent until the opposite is proved) No change in the position of the postulate within the Euclidean system (innocent until the opposite is proved)

(4) *Reductio* attempts to prove that the postulate is a theorem (Saccheri)	*Unsuccessful*: Unsettled fluctuations in the attributed meaning of the postulate No change in its truth value (innocent until the opposite is proved) No change in the position of the postulate within the Euclidean system (innocent until the opposite is proved)
(5) *Reductio* attempts that do not reach contradiction (Saccheri (effectively), Lambert, Gauss)	*Successful*: First grounds to argue that something has changed in the geometrical attitude with respect to the postulate and its geometry. Modified meaning Unsupported by *reductio* truth-value No change in the position within the Euclidean system Preliminary signs for the existence of other geometrical systems, with respect to which the postulate should (eventually) be evaluated
(6) Specific attempts to deny the postulate that lead to hyperbolic type of non-Euclidean geometry (Bolyai, Lobachevski)	*Successful*: Leads to acceptance of hyperbolic counterpart of the postulate and to the existence of a new geometry, based on the rest of the Euclidean axioms and the counterpart Modified meaning: in a new, broader context, the postulate is rivaled by its hyperbolic counterpart Truth-value now is effectively restricted to the Euclidean system and not to *all* geometrical propositions (since Euclidean geometry is not the sole geometry any more) No change in the position within the Euclidean system proper

	Existence of other, non-Euclidean geometrical systems, with respect to which the postulate should be evaluated (false in hyperbolic non-Euclidean geometry)
(7) Specific attempts to deny the postulate that lead to elliptic type of non-Euclidean geometry (Riemann)	*Successful*: Leads to acceptance of elliptic counterpart of the postulate and to the existence of a new geometry, based on the rest of the Euclidean axioms (with exception of the first postulate according to which two points determine a line uniquely). Modified meaning: in a new, broader context the postulate is rivaled by its elliptic counterpart. Truth-value relativized to the Euclidean system strictly and not to all geometrical propositions (since Euclidean geometry is not the sole geometry any more) No change in the position within the Euclidean system proper Existence of other, non-Euclidean geometrical systems, with respect to which the postulate should be evaluated (false in elliptic non-Euclidean geometry; true as a special case in the Riemannian geometry of variable curvature)
(8) Acceptance that the new non-Euclidean geometries are as consistent as the Euclidean one (Beltrami, Klein)	Final of the change: Fifth postulate has a restricted special meaning, preserved within the special case of Euclidean geometry Fifth postulate does not hold for all geometrical cases so it is false in the universality of its original claims (to hold for *all* lines, angles and in *all* cases) Truth-value of the fifth postulate is changed from T to F with respect to the universality of its original claim Truth-value of the fifth postulate is preserved the same (T) within the special case of the Euclidean system

	Position of the postulate is preserved the same within the special case of the Euclidean system Position of the postulate is changed within the meta system of all available geometries (it is not a postulate in all systems but only in the Euclidean system)

Table 5-2

From a formal point of view the revision might have concerned the fifth postulate by itself or the fifth postulate within the system of the axioms of the Euclidean geometry. Let us consider the historical cases with respect to the formal point. If we accept as the first stage of the actual revision stage (6) when for the first time the possibility for the existence of geometrical axiomatic system different from the Euclidean one was recognized, and, as a final stage stage (8), when the alternative geometrical systems were *de facto* established as as consistent as the Euclidean system, then we can investigate the formal point of the revision specifically. During stage (6) the original universal formulation (OF) of the fifth postulate was undoubtedly revised in both meaning and truth-value. For obviously there were lines in the realm of geometry that do not meet at the side of the two interior angles no matter how indefinitely produced as in the case of hyperbolic geometry. Or, there were also cases where there were more than merely one line which turn out to be parallel to a given line that pass through a point not on the given line, as in the case of elliptic geometry. The meaning of the postulate changed with respect to the scope of its validity. The OF clearly meant all lines, all (interior) angles, all points and planes. The new revised meaning of the postulate, however, restricted its scope to *only those* lines, points, angles and planes that are formulated within a geometrical space with zero constant curvature.[13] Given the fact that (before 18th century) no geometrical objects were distinguished with respect to properties of geometrical space like its curvature, the meaning of the terms, employed in the fifth postulate, changed for they now meant not *all* lines, planes, points, etc. but only those, which have the property of being formulated within a space with zero constant curvature.

The truth context of the OF of the postulate has also changed. For the truth of a proposition is in part a function of its meaning and any change or shift in the meaning of the proposition inevitably drags new truth evaluation of the proposition. Before the revision the OF was regarded as universally, intuitively and necessarily true. Not a single historical case of an attempt to revise the postulate before Gauss, Bolyai and Lobachevski had diminished confidence in the truth of the postulate. All the uncertainty regarding the postulate during the millennia of its attempted proof concerned its type as a proposition (whether it is an axiom or a theorem) and never its truth. The postulate was always regarded as a paradigmatic example of a true proposition and this is perhaps best epitomized in Kant's famous role of Euclidean geometry in the Transcendental Aesthetic of the *Critique of Pure Reason*. Therefore, the formulation of the non-Euclidean geometries inflicted a great damage to the authority of both the postulate and the Euclidean geometry by exposing them as not being the only true geometrical propositions but only as being a special case of a much broader geometry. Does this mean that the fifth postulate merely turned from true to false at the time of the formulation of the non-Euclidean geometries? By no means. For the postulate is even today perfectly true in the system of Euclidean geometry. The revision of its assumed truth value was much more complex and subtle and not simply from "true" to "false". But perhaps it would be most adequate to accept that the truth-value "true" of the postulate followed the shift of the meaning of the postulate and "shifted" on its turn. The universal OF, when viewed in the new light of multiple consistent geometrical systems, is undoubtedly false but it seems illegitimate to view the OF outside the system of Euclidean geometry for *the postulate was meant to concern this system only*. In this sense, the multiplication of domains rearranged the truth and meaning of the system in such a way that delineated precisely the correct region of application of the geometrical propositions. Therefore, it does not seem correct to extract the evaluation of the truth of the postulate before the rich system of all available geometries became present. For if we do this then this procedure should be legitimate also for the postulates of the other geometries and in this way we would be forced to pronounce the fifth postulate of the hyperbolic geometry and the fifth postulate of the elliptic geometry as false too when viewed outside of their systems. Certainly, this is not the case, and as far as all geometries are on a par with each other (at least consistency wise) the truth of each proposition should be restricted only to the system of its formulation.

All of the above taken into account, it is clear that the original formulation of the fifth postulate was revised and we can securely accept that the proposition by itself was actually a subject of the revision. From the formal point of view, however, it is interesting to see whether the fifth postulate was *the only object* of the revision. Again, it is clear that historically, the fifth postulate was far from being the only geometrical proposition revised. Apart from the technically necessary modification of the 1st postulate within the system of the elliptic geometry the Euclidean system as a whole suffered from a severe revision in both meaning and truth-value. The process of this horn of the revision might be at best identified as concluded perhaps with the final stage (8) because practically, it is only then when the (relative, as a matter of fact, but that turned out to be sufficient) consistency of the new geometrical systems was established and they became officially true rivals of the Euclidean system. The revision of the system of Euclidean geometry concerned its meaning, its truth and as a new matter of fact, its true domain of application. The Euclidean geometry was not the only geometry to describe geometrical problems anymore, and true as it is for its domain of application, it is not the only true one; there were new rivals to pretend the geometrical crown for correct description of the physical world. The meaning of the Euclidean propositions was not for all points, angles, lines and planes but only for some of them; this represents an aspect of the revision of the meaning of the system. The truth of the Euclidean system was not an absolute geometrical truth as it was until 18th century but it was restricted only to the new domain of its application. The domain of its application now concerned only a portion of the whole new body of geometry. In this way we have more than good reasons to accept that the fifth postulate of the Euclidean system was not the only subject of the revision but rather, the revision affected the whole system of geometry. The crucial role of the seemingly innocent attempts to expose an axiom as a disguised theorem is clearly visible: the revision of the fifth postulate lead to a revision of the whole system of Euclidean geometry. In this way we are presented with a historically unique step ahead of the mathematical knowledge which revealed new unexplored territories and which was triggered by a simple, to a great extent aesthetic unsatisfaction with the apparent complexity of a single geometrical proposition.

The Revision of the Postulate in the Light of the Pure-Applied Geometry Distinction

The fifth postulate says that "if a straight line falling on two straight lines makes the interior angles on the same side less than two right angles, the two straight lines, if produced indefinitely, meet on that side of which are the angles less than the two right angles" *for all straight lines and interior angles*. It is not difficult to see that the universality claim is not simply a luxury but a *conditio sine qua non*: on the one hand, the postulate is a postulate within the only geometry available, that is, the geometry that has the explanatory monopole over all geometrical problems. On the other hand, the postulate would not service well its fundamental function within the Euclidean system as a first principle if it addresses "some" straight lines rather than "all" and "some" angles rather than all. For *if there were* cases where some lines and some angles did not do as prescribed by the postulate it would have had a very restricted application within the system and should not have been included among the most abstract principles of the geometry. Rather, it would have been a locally true proposition and this is not a property an axiom could have. In this sense it is most natural and obligatory that the postulate has a universal pretense over all possible relevant cases.

The universality claim perhaps marks the source of all evil for the subsequent history of suspicion and exposing. For since the time of Euclid and well until the 18th century the propositions of geometry were not merely taken as valid about the points, the lines, the angles and the planes, the actual terms that figure in them. They were taken as a valid description for the physical world as well: the combined "all" of all abstract geometrical cases and all non-abstract physical world cases is undoubtedly a bigger "all" than the "all" of all geometrical cases alone. The postulate was taken as holding for every physical object that appears to have the properties of a geometrical line or at least approximates them sufficiently well for pragmatic purposes. Therefore, an analysis that attempts to clarify the actual revision of the postulate and especially its epistemic aspect should account for the distinction about its domain of validity. There are three possible revision scenarios:

1. The subject of revision is the abstract fifth postulate which holds only for the abstract domain of geometrical objects (points, lines, planes, angles)

2. The subject of revision is the applied fifth postulate which holds for all cases in the physical world

3. The subject of the revision is the "joint" fifth postulate which holds for both the abstract domain and the physical domain

There is one question which bears heavily on the problem of the truth and the meaning of the postulate and this is the question about the orientation of its applicability with respect to the domains of the abstract mathematical objects and the non-abstract objects from the physical world. We have to distinguish between a *precise meaning* of the propositions of geometry and a *common sense meaning*. The common sense meaning might be that because lines, points and planes are so well recognizable in physical objects around us and in some features of them that actually the propositions of geometry are *about* those objects and their geometrical features. For example, the claim that the sum of the internal angles of a triangle adds up to 180 degrees could be illustrated by a typical triangular IKEA coffee table. The triangular piece of wood of the table is taken to be described by the geometrical proposition without taking into account the details of dimensionality (the piece of wood is actually three dimensional and a triangle in Euclidean plane is two dimensional) and geometrical imperfections (when measured with sufficient precision the angles would turn out not to be added exactly to 180 degrees for they are not ideally sharp, etc.). The adequate meaning of the geometrical propositions, however, should do justice to their precise mathematical meaning. In this case the claims about planes, lines and points are claims about planes, lines and points and not about actually observed embedded or reflected geometrical properties of physical objects. For geometrical planes, points and lines are ideal constructions whose precise definitions which provide their mathematical meaning could never be exemplified with mathematically sufficient accuracy in the physical world: for points are zero dimensional and nothing in the physical world is actually physically zero dimensional (zero dimensional should automatically lead to a claim affirming the physical inexistence of such objects and that is not good for the geometrically fundamental object "point" which, above all, needs existence in some sense in order to figure in the system of geometry), lines are one dimensional and stay for an idealized pure length whereas in the physical world nothing is purely one-dimensional but it always has some "thickness" and "width" as well, etc.

Each of the meanings comes with some sort of semantics. The common sense semantics would render as meaningful and true geometrical propositions about chairs and tables whereas a precise geometrical semantics should ensure that the precise meaning of the geometrical propositions is taken into account when they are interpreted with respect to some theory of truth. Which one of those meanings is the adequate one when it comes to the revision of the fifth postulate and the system of Euclidean geometry? This is a question of leading importance for the epistemologist. Historically and especially after the formulation of non-Euclidean geometries people started to regard the question of the "true geometry of (physical) space" as an empirical one, that is, as one that could be only empirically decided. Too often this question has been mistakenly taken as a substitute for the question about the epistemic nature of the geometry itself. For the sake of epistemic precision we have to distinguish between a *revision of the abstract sense of the fifth postulate* and a *revision of the physical sense of the fifth postulate*. Thus, one might argue that two distinct epistemic types of revisions actually took place along the 18th century development of geometry. The more naturalistically minded philosophers would happily accept that both the physical sense of the postulate and the abstract sense of the postulate got revised empirically. The more rationalistic minded ones would accept that even if the physical postulate got revised empirically this by no means establishes that the purely mathematical postulate got revised empirically as well. Arguments that assess the possible epistemic scenarios for the revision of the fifth postulate would be given in the following section.

The distinction between abstract (or pure) geometry and physical (or applied) geometry employed here is a modern one and leads its source from the beginning of the 20th century. It accepts that there is a fundamental difference between the intended domains of the propositions of geometry proper, which employ abstract terms and the propositions of applied geometry, which employs physical terms (interestingly, besides abstract ones). Most notably Rudolf Carnap, in the preface to perhaps the most influential modern book on the problems of space and time, Reichenbach's "Philosophy of Space and Time", writes:

> "It is necessary to distinguish between pure or mathematical geometry and physical geometry. The statements of pure geometry hold logically, but they deal only with abstract structures and say nothing about physical space. Physical geometry describes the structure of physical space; it is a part of physics. The validity of its statements is to be established empirically – as it has to be in any part of physics – after rules for measuring the magnitudes involved, especially length, have been stated."[14]

Carnap uses the distinction with an epistemic purpose: he tries to settle down the complex debate about the alleged synthetic apriority of the propositions of Euclidean geometry which has its beginning in the Transcendental Aesthetic in Kant's *Critique of Pure Reason.*[15] No matter the exact historical origins of the explication of the distinction between pure and applied geometry it is difficult to find reasons to ignore it. For on any theory of meaning that might be of some philosophical interest the meaning of the proposition "Given a line and point not on the line, it is possible to draw exactly one line through the given point parallel to the line" and the proposition "The planets in the solar system move in elliptical orbits around the sun" should be read as being of a different kind. The first proposition employs terms that do not refer to physical objects but to objects at best recognizable as abstract, that is, as being outside space and time and being causally inert. Whereas the second proposition employs concepts that refer both to physical objects (the planets of the solar system) as well as to abstract ones (ellipse). In this sense unless we are willing to accept that the truth conditions of propositions employing abstract terms and the truth conditions of propositions employing non-abstract terms are the same we are forced to accept that the truth-conditions of pure and applied geometry are different enough to be of philosophical interest. The difference in truth-semantics is a sufficiently good reason to introduce a distinction and this is why the pure-applied geometry distinction is a philosophically interesting one. The revision scenarios that account for this distinction have an unambiguous appeal. The validity of geometry for the physical realm is clearly an applicative rather than primary validity. The application of geometry is a function of something, which has already existed as such; for geometry does not need to be applied in order to be geometry in the first place.

The above taken into account, before we clarify the actual epistemic kind of the revision we have to settle the question about which version exactly of the postulate got revised: the *abstract mathematical* one according to which the postulate holds universally for only abstract objects like points, lines and planes, or, the *applied version* where the postulate holds for both the abstract and the physical realm. For the other possible version, where the proposition that got revised is the postulate which referred only to the physical realm is a no go option. In order to accept *a purely physical* meaning of the postulate and no abstract meaning at all we have to agree that the postulate does not concern the entities, whatever they turn out to be, denoted by the terms it is formulated around like “lines”, “angles”, “planes” etc. Those entities, however, are anything but physical entities: they are not to be found in the physical realm and they cannot be observed or measured the way we actually observe and measure the physical objects of our ordinary experience. Thus, the last formulation would be an ultimately bizarre choice and especially so from a semantic point of view; for a proposition naturally is a function of the terms it employs and their meaning. To say that P deals with entities or objects not denoted by its terms but in some other rather exotic way would be clearly beyond reasonable comprehension.

Another remark is worth making with respect to the third possible formulation. One might argue that the entities denoted by the term “line”, “plane” and “point” are actually physical entities and thus they are opened for standard observation and measurement. Or, in a more modest formulation of it, the entities, be they what they may, do have a physical aspect which allows their perception. Recently similar version has been proposed by Penelope Maddy[16] who argued that mathematical entities like sets, for example, are observable in a sort of perceptual sense. The thing is, however, that on any reasonable semantics “line” is only what it means and the definition of "line" is quite explicitly given in the *Elements* under definition 2: "A line is a length without breadth". Besides the trivial fact that no physical property of the line is defined in this or some other definition in the *Elements* the true problem before the purely physical formulation is clearly visible: the line is identified with what usually we refer to as "property" proper: being "long" is not an object in the world, it is a property that we *prescribe to* the object as if it does have it in some kind of way. There is no pure length in the physical world, at least not such which might be observed or measured empirically. For if empiricism is the main source of knowledge about the things in the physical world and no empiricist would easily give this up, it should not have failed by now to perceptually or in some other technologically sophisticated way to register "pure lengths", "pure breadths" and "points" (which seem to be even more difficult to perceive for if the line still has some kind of dimensional property like "being long" the points are deprived even from this last resort of dimensional hope to figure in the physical world where things are usually three dimensional, for they are "that of which there is no part" (*Elements*, Def. 1)). It is not at all clear how the identity of the lines and planes could turn out to be the same as the identity of some empirical counterparts of theirs in the physical world. It is clear, though, that what geometry means when it deals with lines and points *is not the same* with what, say, geology means when measures a plane-like football field with white lines and white dots on it. For theorems of geometry would not have been proven with the intended and the necessary precision if the entities defined by them had some properties actually unaccounted for in the geometrical system. In the case with the purely physical meaning of geometry the truth about, say, the degrees of an angle, would have to be established by the empirical method, through pure measuring and in some cases, like (Euclidean) theorems which claim that *for all triangles* the sum of the interior angles is 180 degrees, it would be quite a challenge to measure all of them in order to get the theorem. The fact that we can meaningfully talk about physical objects having length and angles is quite different from the philosophical question about the true meaning of the geometrical terms and the metaphysical nature of the entities allegedly denoted by them. The system of fundamental definitions and principles, the whole of the definitions, the postulates and the axioms define the true nature of the things the theorems are about. And in this system there is nothing even suggesting that any of the intended objects of the theorems could be of a physical nature; Euclidean geometry is simply not that. The pragmatic use of an abstract system does not change the nature of the system: the fact that I am using the number '2' to ask for a couple of apples does not amount to me perceiving the number. The extreme view that geometrical objects are actually physical entities does not seem to bear under scrutiny; all this goes well with the prevailing attitude in the literature that the objects of mathematics, whatever the solution of the problem of their existence and their properties, are abstract, that is, non-physical entities.

Yet one other option is to say that the intuitive grasp of the truths of geometry is in fact a human perceptual experience and thus we still know the propositions empirically. This goes well with the platonic interpretation of the nature of mathematical objects according to which the referents of the mathematical terms are abstract entities that exist in a platonic realm. The problem of their knowledge is usually solved by accepting that mathematicians have some kind of intuitive intellectual grasp of the truths of mathematics. The option here says that this grasp actually *is* a kind of empirical perception and thus the knowledge about the truths of mathematics and geometry is empirically based. This option, interesting as it sounds, also does not seem to have strength. For it is quite uncontroversial that all empirical observation rests on some sort of causal relation between the observer and the observed entity and this could simply not be the case in the case with the platonic grasp because the realm of the entities is an abstract one and thus they are outside space and time, effectively being causally inert. Consequently, no causal access to them, at least of the standard empirical kind, is conceivable. All those considerations taken into account the revision scenario where the meaning of the fifth postulate is purely physical does not seem very promising.

The other possibility is the revision of the abstract and the applied meaning versions of the postulate. Historically, it is clear that the propositions of Euclidean geometry were taken as valid not merely for the abstract entities "lines" and "points" but also for the things in the physical world. Actually, part (and perhaps not a small part at all) of the confidence in Euclidean geometry stemmed from the fact that when coordinated with actual physical cases the geometrical propositions delivered really well. However, it is important to note that the appeal of the physical validity of geometry is a result not merely due to the contemplation of the propositions but of the reality check that attested that the things prescribed by the propositions are indeed as they said them to be. This, however, demonstrates that there exist some sort of relation between the propositions of geometry and the observed reality. Hans Reichenbach elaborates perhaps most clearly on the nature of this relation, which he calls "coordination", in his seminal book *The Philosophy of Space and Time*.[17] For the present purposes it is important only to note that obviously in the case of applied geometry some coordination between the propositions of pure geometry and the things from the physical world does exists. For otherwise it is not clear how it does happen that, say, the rectangular plot of the table is coordinated with the notion of quadrangle and not, say, a circle or a triangle; illustrations of this kind are countless. In order for the purely geometrical propositions to work recognizably with respect to the physical world they need to be coordinated with the intended objects of investigation and this coordination should naturally follow some sort of rules. Those rules are not part of the purely geometrical system for they are not problems of geometry; we would not find them either in the old book of the *Elements* or in the modern formulations of Euclidean geometry like the Hilbert's one.

In this sense, the abstract fifth postulate and the physically applied fifth postulate are one and the same proposition with the only difference that the latter is coordinated with some objects from the physical world and thus provides one more thing to be revised, namely, the *applied claim* that the selected portion of the world is geometrically indeed as the purely geometrical fifth postulate says. In this way there is one more parameter of the fifth postulate to be inquired about, its *applied validity*. This inquiry, however, is not a purely geometrical one for it concerns objects, referred to by no terms in the original abstract postulate. They refer to, say, guitar strings or ropes and the original fifth postulate is not about these kinds of things; it is about lines and strings and ropes which are certainly not lines for, among other things, they have breadth and, according to definition 2 in the *Elements*, lines do not have breadth (interestingly, their breadth is not negligible but monstrous under a microscope and this sort of observation is what physics usually does for it delivers the best insight into the nature of the object). In this way it is not difficult to see that a revision of the meaning of the pure postulate would automatically lead to a revision of the meaning of the applied postulate but not vise versa. Because if, for example, a case of coordination of the fifth postulate with some sort of physical state of affairs fails, this could happen to be due to a lousy coordination and not simply due to purely geometrical inadequacy of the postulate to be coordinated. Also, one might argue, that a failed case of application of the postulate when it is exposed through a physical counterexample should serve as a reason to revise not merely the applied postulate situation (which consists in the pure postulate, the physical state of affairs and the coordination between the two) but the purely geometrical one.

The problem of the empirical revision of mathematical statements is complex enough and I certainly cannot hope to resolve it here. It is worth noting, however, that a revision of a proposition could only be done if its intended domain is not in the way the proposition says it to be. Translated for the purely geometrical postulate this should mean that the situation with the lines, planes and points is not the one described by the postulate. This, however, is difficult to be delivered through an applied postulate revision. For none of the terms of the revision would be completely stripped of its physical clothing and of its way of coordination as the original terms of the postulate are. The fact that a guitar string does not cross a rope when produced indefinitely (whatever this might happen to mean for physical objects which are pretty much finite in length) says nothing about the behavior of lines; again, for strings and ropes are not lines and naturally they differ in their properties. This difference in property does not permit an automatic conclusion to be transferred from the applied case to the pure one; it simply does not deal with the proper terms. In order for a revision to be a revision of the pure postulate it would have to concern only the pure geometrical terms, their intended referents and the relations between the terms and the referents. As far as no physical object is or could be a referent of purely geometrical proposition new information about the way some physical objects are is irrelevant for the statement of the pure proposition. In this sense, the revision that is of interest for the present purposes is not a revision of any case of failed application of the fifth postulate; very much on the contrary. The whole history of mathematicians trying to prove the postulate, even if carrying some load of applied functionality, shows that it was the abstract meaning that suffered the increased interest. None of the revisers ever doubted the truth of the postulate. If it were the case that empirical consideration lead to the revision first of all, those considerations should be demonstrated. And this is not an easy thing to do since not until the beginning of the 20th century physical illustrations were available that showed the validity of non-Euclidean propositions. And it is quite difficult to see how an empirical situation that complied well with the Euclidean picture (before the beginning of the 20th century) could have been used to revise the very system it complies with.

All of these considerations show that we have to distinguish the pure geometrical revision and the applied geometrical revision as being of different kinds. Whereas the applied geometrical revision might perfectly well be an empirical one due to the reality check that establishes the truth of the conjucted pure postulate plus its coordination, regarding the coordinated physical state of affairs the pure revision is not open for an empirical check of this type. The main reason is that the intended domain of validity of the pure postulate does not have among its members empirically observable entities. And as far as it is this domain that decides the truth or falsity of the proposition, as well as its meaning shifts, empirically imposed probing of the domain is doomed to be fruitless. Besides this the historical case of the revision shows that it was a mathematical and not physical revision. The mathematicians were concerned with the function of the fifth postulate within the system of Euclidean geometry and not with the degree of confidence in its application to the physical world; this confidence was never historically disputed until the beginning of 19th century when the young Bolyai argued that the question about the true geometrical properties of the physical universe could not be resolved through mathematical reasoning alone.[18] Even then the applied validity of the fifth postulate was strictly not disputed: for no empirical counterexamples were available that conformed to the non-Euclidean geometries.

The physics needed some more time until it managed to come up with such counterexamples. As a consequence, no empirical revision might have come to the applied fifth postulate until the beginning of the 20th century and therefore, it seems reasonable to accept that even the applied postulate has not been empirically revised during the quest for the proof of the pure proposition. The main revision, which represents the current illustration of a historical a priori revision is the last option left: the mathematical revision of the pure fifth postulate.

The Epistemic Type of the Revision

Epistemology deals, among other things, with the problem of epistemic kinds. The epistemic kinds are supposed to provide useful information about the nature of justification and knowledge. History of philosophy, through a kind of epistemic natural selection filter, has produced two main competing epistemic kinds: the epistemic kind of a posteriori (or empirical) and the epistemic kind of a priori (or non-empirical). The appeal behind this distinction is not too difficult to identify. We learn a great deal about things through experience and thus it seems quite natural to regard experience as a source of justification and knowledge. Since, more or less obviously, our justifications and the knowledge they sometimes manage to lead to, is not *all the time* empirical, paradigmatic cases here would be knowledge of mathematics and logic, we need to regard the source of this knowledge, whatever it turns out to be, as well. Obviously, we cannot regard it as identical to the empirical one and this naturally produces the other term of the distinction, the a priori one. Historically, a lot of attempts have been made to argue that the very term and consequently, the notion is too vaguely defined in order to be philosophically useful and the debates continue as we speak. However, similar sorts of arguments have been raised against the empirical kind as well. It is beyond the present purposes to engage in comments about the vitality of the a priori – a posteriori distinction; rather, it should suffice to acknowledge that it seems useful enough for anyone who accepts the empirical kind as philosophically meaningful to explore its possible alternatives with the same degree of epistemic interest.

In order to establish the epistemic kind of a justification one needs to identify the epistemic nature of one's reasons to hold a proposition P. Thus if Susan holds that the color of the big spot on Jupiter is red (P) in color, Susan (most probably) has some kind of observational reasons to do so. She, being an astronomer, has a good telescope which permits her to acquire information about the color of the spot. Thus, fallible as they are, Susan's reasons to hold that P are of empirical character. Let us suppose now that Susan is making a collection of propositions that further concern the spot, the construction of her telescope and the nature of color (P_1, P_2, P_3, P_4, P_5, etc). Those propositions also have some kind of reasons to be held by Susan and also, they have some epistemic nature. Let us suppose that the empirical nature of all the individual reasons behind holding the propositions is of the same empirical kind as the ones behind P:

Epistemic reasons to hold that P = empirical (E)
Epistemic reasons to hold that P_1 = empirical (E)
Epistemic reasons to hold that P_2 = empirical (E)
Epistemic reasons to hold that P_3 = empirical (E)
Epistemic reasons to hold that P_4 = empirical (E)
Epistemic reasons to hold that P_5 = empirical (E)

Now, let us imagine that between the set of propositions ($P_1 – P_5$) and the proposition P some logical tension appears of the kind that S cannot both hold P and ($P_1 – P_5$); say, all the numbered Ps in the set go really well with each other but they do not go well with P. Thus Susan, being a rigorous astronomer, cannot leave the things like that and has to do something in order to get rid of the tension. What she would do is to examine both the source and the strength of her reasons to hold that P and the individual Ps in the set ($P_1 – P_5$). Also, for the numbered Ps come in some sort of a set, she would assess the source and the strength of her reasons to hold this particular set of Ps. At the end she would reassess the source and the strength of her reasons to hold that P. Eventually, she would find some of the reasons weaker than some of the others and thus she would revise her attitude towards the whole set ($P – P_5$). Say, she finds her reasons to hold that P weaker than her reasons to hold each and individually any (or some, depending on the case) of the numbered P's as well as the set (P_1 – P_5). Also, say, she discovers that the telescope has a malfunction to present under some particular conditions brown colors as red. Then, Susan would substitute P with P*, a proposition which meaning shifted in such a way that goes well both with her reasons to hold it and her reasons to hold the set (P_1 – P_5) or with Q, a proposition which is so different in meaning from P that cannot be said to be a modified version of it. At the end of the day she would have a new set of propositions that concern the Jupiter spot. The new property of this set as compared to the old one is that the logical tension between the propositions is eliminated and her reasons to hold its individual propositions as well as the set itself are stronger than the previous ones. In this simplified scenario the astronomer *revises* a scientific belief about a particular state of affairs of the physical world. Can we inquire about the epistemic nature of the revision? Does the revision have an epistemic nature of its own, rather than being just a bunch of unconnected epistemically kinded reasons to hold the individual propositions? Well, for one thing, the revision took place for a reason and in this sense the revision does have its own justification. Naturally enough, the justification might and perhaps should consists (probably partially) of the constituent individual reasons to hold the individual propositions involved in the procedure. Yet, the reason is different from every one of the individual ones and thus, it is sufficiently well defined to be inquired about. It is the task of the epistemologist to identify its epistemic kind. Now, what could possibly be the epistemic kind of a revision of this type where all the propositions involved were empirically justified? Is it conceivable that all the constituent propositions of the revision are empirically revised, the revision follows the rules of the working revisionary logic and yet the final result is to be regarded as non-empirically justified? By all epistemic standards it does not seem so. For the epistemic kind of the reasons to hold that P (or set of Ps) and the end product of the revision is a P (or set of Ps) and is identified with respect to the actual source of the reasons. That is, it must be established whether the reasons, whatever they are, come from experience or not. In the case of Susan's revision all the constituent reasons are empirical and if we accept that the jump from a set of premises to a conclusion (the result of the revision) is governed by experience (as it usually happens in science) there is no other option for the reasons that hold the conclusion but to be regarded as empirically justified as well:

1. Epistemic reasons to hold that P = empirical (E1)
2. Epistemic reasons to hold that P_1 = P_5 empirical (E2)

3. Tension between P and $P_1 = P_5$
4. Justifications evaluated empirically
5. E1 empirically abandoned in favor of E3 (reasons to hold that P* or Q)
6. Epistemic (E) reasons to hold that ($P - P_5$) abandoned in favor of epistemic (E) reasons to hold that ($P^*(Q) - P_5$)
7. Result of the revision: ($P^*(Q) - P_5$) + the epistemic reasons (E) to hold it

Thus, it seems quite adequate to accept that revisions do have their own epistemic kind and second, that this kind is traceable to the constituent epistemic kinds of the justifications of the propositions involved in the procedure of revision. Now, the following reasoning naturally imposes itself: if there is a general epistemic form for a revision and we are presented with actual historical examples of empirical versions of it what about non-empirical versions, that is, what about a priori ones? Is it not the case that it is philosophically interesting to inquire about revisions with the alternative epistemic kind? Intuitive feeling for logical symmetry tells us that it naturally is. Given the prevailing view that mathematics and logic are indeed domains where knowledge is delivered through empirically independent means this question seems even more natural. For mathematical and logical knowledge do change and consequently, it is quite natural to suppose that the revisions in mathematics and logic do have an epistemic aspect of their own.

In order to establish the epistemic kind of the revision of the fifth postulate we need to identify the participant propositions and the epistemic kinds of their justification. Now, the justification here would be used in a sufficiently loose way as "reasons to hold that P" that would allow to use it not only in the case with the theorems where the proof is a kind of logical reason to accept the theorem, but also to accept the axioms which are not proved. The objection which claims that the axioms and postulate (as well as the definitions) are not to be assessed with the standard of "truth" and "falsity" for they are propositions accepted in a sort of a conventional way seems to be an interesting one. It is certainly in the nature of the fundamental principles of geometry not to be proved logically but just to provide a fundament that would allow all the proofs down the geometrical road. This, however, by far does not mean that the choice of the axioms is arbitrary and virtually many rival propositions might have been fit in their place. For there is a very clear indication that decides the choice of an axiom or a postulate as a good one or bad one: the properties of the system that flows out of them. If the choice of the axioms is such that the system of theorems proved from them is too weak, that is, it is not very useful for coping with a large range of problems, then the choice of at least some of the axioms is not very good. If from the chosen axioms propositions are proved that contradict each other then the resulting system is an inconsistent one and the choice is definitely a bad one. Similar sorts of reasons reveal that in fact the choice of the axioms is a pretty delicate problem and in this sense, we can quite sensibly distinguish between "good" axioms and "bad" axioms; this points against the arbitrariness portion of the objection.

The case with the inconsistent system is also instructive for the ability of the axioms to be assessed by the notions of "truth" and "falsity". For even if being held on different grounds than the theorems, the axioms do have a definite say about the subject matter of, in this case, geometry. And therefore they could be assessed in exactly the same way as the theorems: by simple check whether what the postulate affirms about the subject matter is like it says or no. Thus, for example, if the fifth postulate of the Euclidean system were to claim that there are exactly 5 lines that go through a point not on a given line that are parallel to that line, it would not have passed too much time until the postulate is actually exposed as false and simply as an unsuccessful convention. For it is clear that what the postulate is saying is not the case. In this sense, the first principles of a system are as approachable with the notions of "truth" and "falsity" as the rest of the system. It is very true, however, that some of them look truly difficult to assess as being true or false. To give another example let is take the postulates 1 – 3 which concern geometrical procedures rather than factual claims about its subject matter. How could the postulate that says "to produce a straight line from any point to any point" could be assessed as true or false? Well, obviously not so intuitively easy as, for example, postulate 4, which says that all right angles are equal to each other. The former looks much more like a rule than as a proposition and in this sense it is not at all clear how it could be examined with respect to a factuality of its claim, the usual mechanism which establishes truth or falsity.

One way to look at this problem is to accept the proposition as making a modal claim that *it is possible to produce a straight line from any point to any point*. In this case the postulate has a statement that is factual enough to be truth-interpreted. For now the conditions under which the claim would not be true are pretty clearly stated: if there were a case where we actually cannot draw a straight line from a point to another that would show that what the postulate is affirming is not the case and thus it could be accepted as false. Another way to look at it is to look at the system resulting from the choice of the postulate: if there are theorems proved that contradict each other (and this somehow is traceable back to exactly this postulate) then, it could be said that because the postulate leads to an inconsistent system it does not contribute to its truth and therefore, it is as good as false. One way or another, the issue of truth and falsity of axioms is certainly too complex to be examined in sufficient detail here. What is sufficient for our purposes is to acknowledge not the truth or the falsity of the axioms of Euclidean geometry but the fact that they figure as axioms and postulates *in a justified manner*. In this way the epistemic talk with be cashed out in terms of justification and not in terms of truth which is much more secure from an epistemic point of view; recently such style of analysis is particularly popular among epistemologists; again, Albert Casullo's recent book on a priori justification[19] provides a good illustration of that. Thus, we can examine the epistemic nature of the reasons to hold (or accept) the propositions participating in the revision of the postulate and not engage in arguments about their truth or falsity.

If we ignore the directions of the historical attempts to prove the postulate that did not lead to success, namely, the formulation of geometry as consistent as the Euclidean one, we can extract the following structure of the successful revision:

1. Fifth postulate of Euclidean geometry (pure abstract meaning)
2. Acknowledged suspicious complexity of the postulate
3. Attempts to prove the postulate and to show that it is a theorem
4. *Reductio ad absurdum* attempts (Gauss, Bolyaj, Lobachevski)
5. *Reductio* does not reach contradiction
6. Acknowledged the existence of alternative geometries (hyperbolic and elliptic)
7. Acknowledged the existence of alternatives of the original fifth postulate (hyperbolic postulate and elliptic postulate
8. Acknowledged the (relative) consistency of the new geometrical systems
9. The fifth postulate receives restricted meaning, validity and truth in the Euclidean system which is now just one of the many possible geometries

Each of those steps has to be assessed epistemically in order to identify the epistemic kind of the revision. Historically, the fifth postulate was held to be intuitively clear and compelling. In fact Kant, as late as late 18th century, influentially regarded the propositions of the Euclidean geometry as true with necessity and apodictic certainty.[20] The intuitive appeal of the statement of the postulate is certain in its role to be a reason to actually hold the postulate as a first principle of the geometry. If we use modern terminology, in order the postulate to be held not even compellingly true but simply true, there should have been an experience that corresponds to the proposition. This experience should have had the following characteristics, among others: it should have provided clear cut evidence for the behavior of straight lines for all lines, angles and points. Also, it should have done it in a way that would not allow the imagination of an alternative. At the end, the experience should have concerned precisely and adequately the terms, employed in the proposition and in the case of pure geometry these are terms that refer at best to abstract referents. Given the fact that, as we have seen in the above sections, no experience could relate to an abstract referent the last requirement fails to obtain. Further, no experience has the property of delivering true knowledge with necessity. For it always relies on contingent means of delivering the information and hence all knowledge, which is based on experience, is contingent, that is, the described state of affairs could well happen not to be the case; this feature of empirical justification and knowledge is recognized by the overwhelming majority of empiricists. Therefore, it is practically impossible to see how contingently delivered knowledge could have at the same time the property of being true with obvious necessity; for it always contains within itself the (sometimes implicit) notion that it could nevertheless be otherwise. And yet the fifth postulate was held not merely historically but also epistemically on modal grounds; it has been viewed as necessarily true and the imagination of the possibility to extend a straight line on the side where the sum of the interior angles is less than 180 degrees sufficiently far and not actually crossing the other line was considered impossible. As far as necessity could not be delivered via empirical observation this impossibility of conceiving of an alternative, which is a primary reason to hold the postulate, could not be delivered though experience. Thus the second requirement fails as well. The first requirement could not be met empirically too and for the following reasons. First, there is no adequate and precise counterpart in experience for the terms of pure geometry. There are no points, lines, planes and angles in the physical world and consequently, we cannot perceive them.

In addition, the experience in question should have been able to demonstrate an infinite extension of a line. Even if we somehow are presented with such experience it would have to be visualizatory infinite in order to keep the whole of the lines in the perceptual or visual field at all times, that is, at all infinitely many instances of time, during which the extension takes place. Thus it seems clear that no experience whatsoever could satisfy the infinity request in the postulate; if experience could not provide the grounds for this there must be some other source for it. Another difficulty with the alleged empirical justification of the postulate regarding its portion for the infinite extension involves the universality of its claim. Even if we imagine that a single case of physically infinite extension is actually observed it is not at all clear how this experience would manage to extrapolate over *all* similar cases and even worse, how it would manage to extrapolate with the intuitive clarity and certainty of the situation, depicted in the postulate. At the end of the day, however, all these infinite desiderata are certainly impossible in the physical world, for even if there were a geometrically proper line in it an infinite extension would have been an infinite task that would require an infinite amount of physical space, an infinite amount of energy to conduct and an infinite amount of time to conclude. Being physically impossible, they are, as a consequence, unavailable to the experiencing mind of the agent who supposedly justifies her holding that the fifth postulate. Therefore, the first requirement also fails and thus, by the method of exclusion we are compelled to hold that either the fifth postulate was not justified or, that it was justified independently from experience. Given the obvious fact that there were more than serious reasons to have the fifth postulate among the ten most important geometrical principles, good enough to count as epistemic justification, we have more than good reasons to accept that the fifth postulate was justified a priori.

The second step of the revision is the acknowledged suspicious complexity of the postulate. When compared to the others and to the axioms the fifth postulate looked to the ancient mathematicians too complex to be a fundamental principle. For the fundamental principles were naturally expected to be as simple as possible since any complexity might have turned out to lead back to other propositions, from which the complex one might be deduced. In this way the fifth postulate was deprived from confidence with respect to its formal function within the system of Euclidean geometry: it was not intuitively clear that the postulate is indeed a postulate and not a theorem, a proposition provable from the other axioms. It is important to point out that the postulate was not deprived from confidence about the fact of its truth: and the mathematicians, trying to actually prove the proposition, demonstrated this in a brilliant way. As far as the earliest *reductio ad absurdum* approaches were much later (and which, again, did not target proving the falsity of the postulate but the *impossibility* of its being false) they were looking for other ways than just intuitive and necessary appeal, to justify its statement. In this sense the mathematicians were actually trying to reach an alternative justification for the postulate and this amounts to an attempt to revise the available justification. The recognition, however, of a formal complexity has nothing to do with any empirical observation whatsoever: no observations were made of the physical world, no empirical events suggesting this complexity were observed. The deprived confidence was due solely to a mathematical reflection over the formal properties and the function of the postulate. In this sense, epistemically, the second step was also conducted independently from experience. Historically, no positive evidence is available that empirical considerations actually led people like Ptolemy to acknowledge the complexity on empirical grounds.

The third step contains the huge number of historical attempts to prove the postulate. Whereas none of those actually leads to a revision this step does not have a direct bearing on the actual historical revision. It is quite clear, however, that all these attempts were purely mathematical and relied on available mathematical and logical techniques. No mathematician tries to prove a theorem about mathematical entities going on the field and measuring some properties of the land. What she does typically is that she produces a kind of mental image of the problem and by means of logic, intuition and available geometrical knowledge tries to see what could be used as a mathematically legitimate route to the conclusion, that is, the theorem. In this sense even those steps were non-empirical, albeit they did not contribute logically to the final revision. The main reason why they failed is that none of them employed the logical method of *reductio ad absurdum* which proved to be the successful one, as history of mathematics shows. It is beyond doubt, of course, that all of these contributed significantly to the weight of the problem and the intellectual challenge it posed before the mathematicians.

The fourth step is the crucial one. It involves the actual logical procedure that led to the formulation of the non-Euclidean geometries and in this way it is at the heart of the revision. The procedure consists in the following: the negation of the proposition to be proved is assumed and the logical consequences of this assumption are examined. When a contradiction is reached this is taken as a proof that the negation of the proposition is false, that is, that the proposition is true. Lack of contradiction, strictly logically speaking, proves nothing, for it is always conceivable that the contradiction is still further down the road of the farther logical consequences. However, when compared with the consequential level of another system, whose consistency is accepted (like the one of the Euclidean geometry), the lack of contradiction, if reached at the same level as the lack of contradiction of the relative system, might be sufficient to accept that (relative to the system) no contradiction would be reached. Historically, this proved to be the case. Bolyaj and Lobachevsky negated the fifth postulate and examined the consequences of this negation. Eventually, they reached no contradiction but in the meanwhile, they proved quite a lot of theorems that seemed pretty instructive about the fact that the system they were proved in is in fact a consistent system. Technically, the consistency of the non-Euclidean geometries was proved by Eugenio Beltrami and Felix Klein. However, what they did was not to supply an independent proof of consistency but a proof of *relative* consistency, that positioned the new systems on a par with the Euclidean system. In this way the consistency of the new systems was made dependant on the consistency of the old system: if the old system is consistent so are the new ones. And since the old system was examined for contradictions for more than two millennia by a great deal of good mathematicians its consistency was practically out of the question.[21] Effectively, this amounted to accepting the new systems as consistent ones. From an epistemic point of view the *reductio ad absurdum* is a standard logical procedure for proving. The strength of the proof comes from the unavailability of conflicting propositions deduced on the basis of the negation of the proposition to be proved. Logic, however, is a discipline about the way human thought works and not about how human perception works. It deals with propositions, subjects and predicates, quantifiers and other instruments and it examines, among other things, the transitions from a set of propositions to a proposition not in the set. The main authority of logic is thought itself and it does not need an appeal to actually observed states of affairs in order to be able to judge off the logically legitimate transitions from the logically illegitimate ones. In the case of geometry logic is an indispensable tool for it grants the validity of the mathematical proofs. The interesting thing about the *reductio* of the fifth postulate, however, is the following: it could not have possibly been done by appeal to experience. For the whole available experience until the beginning of the 20th century, when actually Einstein showed that the physical world follows the rules of the non-Euclidean geometry, there was no even a single observed counterexample for the Euclidean propositions. As far as the world was concerned it was undoubtedly a Euclidean one. Surely, there were imperfections in perception, fallacies of perspective and all sorts of perceptual phenomena that might have led one to think that some of the Euclidean propositions were physically not the case. But it inevitably turned out that all these were due not to non-Euclidean properties of the physical objects but due to peculiar features of our perceptual apparatus or the laws of optics that governed the behavior of the light before it gets in the apparatus. Never an optical illusion actually managed to grow as epistemically strong as to put the application of the Euclidean propositions to the physical world under doubt. In this sense, it is as clear as it gets that the revision of the fifth postulate could not have even taken off the ground if it were the case that it was empirically justified. For the very first line, the negation of the postulate (be it the hyperbolic or the elliptic version of the negation) is contradicted by the available experience and consequently, the whole procedure would have been empirically illegitimate right from the start. The only kind of justification that could have allowed the acceptance of the negation of the postulate is the logical and mathematical imaginative reflection. And this is not of an empirical kind but much on the opposite, it could work *in spite of experience*. In this way the revision of the postulate was not merely independent from experience but it was possible *only as far as it is independent from experience*. For the available experience would have not allowed the procedure to even begin, let alone justify some conclusion out of it. Thus, the fourth step, the one that is the moving engine behind the revision, was only possible as independently from experience, that is, in an a priori way. Steps 5, 6 and 7 are complementary to step 4 because they constitute the finalization of the procedure of *reductio ad absurdum*. They follow naturally as part of the reductio and therefore they share the epistemic kind of the whole procedure. No new justification jumps in or substitutes the initial reductio justification and so we can regard them as a priori conducted as well.

Step 8 concerns the properties of the newly discovered systems, namely, the property of their consistency. Beltrami's and Klein's proofs are paradigmatic mathematical proofs and consist in coordination between the propositions of the Euclidean geometry and the propositions of the non-Euclidean ones. In this way they participated in the revision of the meaning of the postulate for now the postulate was officially not absolutely true within the only consistent system available. The acceptance of the new geometries as consistent alternatives of the Euclidean one restrained effectively the domain of application of the fifth postulate (as well as of all other Euclidean propositions): from now on they were valid only for flat space with zero constant curvature and not valid for any other space. From an epistemic point of view the important question here is the justification of the proof for relative consistency of the non-Euclidean geometries. The proof was driven by logic and conducted in a mathematical way. In fact, Beltrami constructed a mathematical model for 2-dimensional non-Euclidean geometry within the system of the 3-dimensional Euclidean geometry.[22] His model concerned the hyperbolic geometry and demonstrated a case where the rest of the postulates held (postulates 1 - 4) but not the fifth one. Later, Klein provided models for other versions of the non-Euclidean geometry including the Riemannian elliptic geometry of positive curvature.[23] The proofs concerned the abstract systems of Euclidean and non-Euclidean geometries and clearly, no empirical observation triggered or made them possible. The problem of consistency of a system is even on a higher abstract level than the mere geometrical problems for it concerns the properties of the system itself and thus it retracts to a meta level of investigation. The coordination between the propositions of one of the systems and the propositions of the other could not have been regulated empirically since the experience would always prefer one of the propositions before the other (for they contradict each other) and so it would misbalance the coordination epistemically. Thus, there are good enough reasons to accept that the relative consistency proof was of the same epistemic kind as all mathematical and logical proofs until now. The last step 9 concludes the revision of the fifth postulate of Euclidean geometry. The postulate begins to function in the now restricted domain of Euclidean geometry and this new function naturally flows from the preceding steps of the revision. The justification for this restriction is the same justification as the one behind steps 5, 6, 7 and 8 and thus an a priori one. In this way we receive the following epistemic picture of the revision of the postulate:

1. Epistemic reasons for holding the Fifth postulate of Euclidean geometry = A priori
2. Epistemic reasons behind the acknowledged suspicious complexity of the postulate = A priori
3. Epistemic reasons behind attempts to prove the postulate = A priori
4. Epistemic reasons behind the *Reductio ad absurdum* attempts = A priori
5. Epistemic reasons behind *Reductio* does not reaching contradiction = A priori
6. Epistemic reasons to acknowledge the existence of alternative geometries = A priori
7. Epistemic reasons behind acknowledging the existence of alternatives of the original fifth postulate (hyperbolic postulate and elliptic postulate) = A priori
8. Epistemic reasons to acknowledged the (relative) consistency of the new geometrical systems = A priori
9. Epistemic reasons behind the fifth postulate receiving restricted meaning, validity and truth in the Euclidean system (now just one of the many possible geometries) = A priori

All of the above taken into account should be sufficient to show that the epistemic kind of the revision of the fifth postulate of Euclidean geometry is the a priori epistemic kind. The revision represents a historically famous growth of the mathematical knowledge which led to the discovery of entirely new mathematical fields. The main regulatory mechanism of this revision was independent from experience and thus serves as a good illustration of the historical potential of empirically independent scientific revisions. The example is particularly instructive in the case of Friedman's spatiotemporal framework within his three-layered model of scientific knowledge for it demonstrates how the dynamics of spatio-geometrical principles, which are at the core of the model, could be a rationalistic and not an empirical one.

What if the revision of the fifth postulate is attacked as being strictly (purely) mathematical and thus, not pertaining to the domain of natural science; hence, accused as being hopelessly irrelevant for science as well as non-informative about the epistemic nature of naturally scientific revisions proper? At the end of the day such an argument does have a certain appeal and especially in the light of the defended here pure-applied geometry distinction. The thing is, however, that the result of the revision was not merely a shift in the meaning of a single geometrical proposition, which, by virtue of its pure mathematical nature is not valid for the realm of the physical world, the traditional domain of the natural science. Not at all, for firstly, even purely mathematical as it is the fifth postulate of Euclidean geometry has been applied for many centuries to the physical world within the system of the applied Euclidean geometry; and secondly, the main result of the revision was the birth of the very tool of the modern natural science – the Riemannian geometry of manifolds. True enough, the pure – applied geometry distinction holds here as well as in the case with the Euclidean geometry but this does not annihilate its applied function. On the opposite. The natural science and physics in particular would collapse without its main instrument of formulation and description of physical phenomena; in case of space this is the Riemannian theory of manifolds which was the most powerful "side result" at the end of the complex history of the fifth postulate revision. In this sense, mathematical as it is, the revision does bear heavily on natural science and mainly so by virtue of its role as applied mathematics. As far as applied mathematics is at the core of the natural science this purely mathematical revision, by simple transitivity, turns out to be indispensable for the natural science itself.

At the end, the following seems like a legitimate question too: even if the described case is actually what historically happened, how useful is this from a philosophical point of view rather than from merely historical one? A mathematician could have all sorts of reasons, not very good ones as well, for that matter, and she could perfectly well reach what could independently turn out to be a valuable piece of scientific knowledge. How we are then to distinguish between epistemically good justifications as opposed to just lucky historical justifications? The epistemic success is, however, perhaps the best parameter for a justification to be assessed with respect with. For the main purpose of a justification is to lead to knowledge and if a justification manages to do so it is epistemically more valuable then a justification which is considered epistemically the state of the art but fails to deliver the goods. In this sense, to take science, for example, in the extreme case when scientists for ages got it right by mere epistemic luck, the epistemologists might have to look more carefully at the red lamp blinking and put under scrutiny their epistemic standards for good and bad as well as the very definition of epistemic luck rather than the actually gained knowledge.

Notes

[1] References to Euclid's Elements follow the translation of Heath, Thomas L. [1956]. *The Thirteen Books of Euclid's Elements* (3 vols.), 2nd ed. [Original publication: Cambridge University Press, 1925], New York: Dover Publications

[2] The intuitive obviousness is well read from the axioms even today, in the presence of many non-Euclidean geometries: The First Axiom says that things which are equal to the same thing are also equal to one another; the Second Axiom says that if equals are added to equals, the wholes are equal; the Third Axiom says that if equals be subtracted from equals, the remainders are equal; the Fourth Axiom says that things which coincide with one another are equal to one another and the Fifth Axiom (not to be confused with the fifth postulate) says that the whole is greater than the part.

[3] Consider: the definition of a point says that a point is what has no parts.

[4] In fact the consistency of Euclidean Geometry was historically never strictly proved until Hillbert reduced the problem of consistency of Euclidean geometry to the problem of the consistency of arithmetic thus providing a proof for the relative consistency of the Euclidean geometry. The main confidence in the consistency of the Euclidean geometry stems mainly from the fact that during the centuries contradicting theorems were never actually found.

[5] Proclus [1970] *A Commentary on the First Book of Euclid's Elements*, Princeton, NJ: Princeton UP.

[6] Saccheri, G. [1733] *Euclides ab omni naevo vindicatus sive conatus geometricus quo stabiliuntur prima ipsa universae geometria principia.* Mediolani: Ex Typographia Pauli Antonii Montani. (Reprint, with facing translation by G.B. Halsted: New York, Chelsea, 1986).

[7] Lambert [1766] *Theory of Parallel lines* (Die Theorie der Parallellinien) Lambert, 1895 J.H. Lambert, Theorie der Parallellinien. In: F. Engel and P. Stackel, Editors, *Die Theorie der Parallillien von Euklid bis auf Gauss*, 1895 - BG Teubner.

[8] Gauss, C.F. – he never published his work on the non-Euclidean geonetries, rumors tell that it was mainly due to Kant's authority at the time

[9] In 1831 he published "Appendix Scientiam Spatii Absolute Veram Exhibens" ("Appendix Explaining the Absolutely True Science of Space"), a complete and consistent system of non-Euclidean geometry as an appendix to his father's book on geometry, *Tentamen Juventutem Studiosam in Elementa Matheseos Purae Introducendi* (1832; "An Attempt to Introduce Studious Youth to the Elements of Pure Mathematics").

[10] Lobatchevsky N.-J. [1835] "Géométrie imaginaire (M. U. Ka., 1, 3-88, 1835 ; Œuvres complètes, réimpression, 1, Kasan, 1883)."
[11] Riemann, B. [1868] "On the hypotheses which lie at the foundation of geometry" in Ewald, William B., ed., 1996. From Kant to Hilbert: A Source Book in the Foundations of Mathematics, 2 vols. Oxford Uni. Press: 652-61.
[12] In 1795 John Playfair published this formulation of the Euclid's parallel postulate called Playfair's axiom although he explicitly denied credit for it.
[13] The notion of curvature of space was historically formulated for a first time by Carl Gauss as a real-valued function (Gaussian curvature) and represents the local deviation of a surface from flatness.
[14] Carnap, R. [1950] *Introduction* to Reichenbach, Hans [1950] *The Philosophy of Space and Time*, Dover, p. vi.
[15] For Kant's exposition see Kant, Immanuel [1781/87] *Critique of Pure Reason*. Trans. Norman Kemp Smith as *Immanuel Kant's Critique of Pure Reason*. London: Macmillan Co. Ltd., 1963.
[16] For details see Maddy, P. [1980] "Perception and Mathematical Intuition" in *The Philosophical Review*, Vol. 89, No. 2. (Apr., 1980), pp. 163-196.
[17] For details on his account of coordination see Reichenbach, Hans [1950] *The Philosophy of Space and Time*, Dover.
[18] Bolyai, Janos [1831] "Appendix Scientiam Spatii Absolute Veram Exhibens" ("Appendix Explaining the Absolutely True Science of Space"), an appendix to his father's book on geometry, *Tentamen Juventutem Studiosam in Elementa Matheseos Purae Introducendi* (1832; "An Attempt to Introduce Studious Youth to the Elements of Pure Mathematics").
[19] Casullo prefers to present his analysis of the a priori in terms of justification and not knowledge. His reasons in a way summarize the prevailing epistemic attitude of preference towards justificatory analysis. For details see his Casullo, Albert [2003] *A priori Justification*, OUP.
[20] Kant, Immanuel [1781/87] *Critique of Pure Reason*. Trans. Norman Kemp Smith as *Immanuel Kant's Critique of Pure Reason*. London: Macmillan Co. Ltd., 1963.
[21] It is worth noting some exceptions of the rule; for example, the notorious proof within Euclid's system that all triangles are isosceles. David Hilbert rectified this in 1900.
[22] Beltrami, Eugenio [1868] *Essay on the Interpretation of non-Euclidean geometry* in Beltrami, Eugenio [1835-1900] Opere matematiche di Eugenio Beltrami / pubblicate per cura della Facolta di scienze della R. Universita di Roma, Milano, U. Hoepli, 1902-1920.
[23] Klein, Felix [1871] "Über die sogenannte Nicht-Euklidische Geometrie", in *Mathematische Annalen*, 4: 573 – 625.

CHAPTER VI

HISTORICAL CASES OF A PRIORI REVISABILITY IN NATURAL SCIENCES: THE REVISION OF THE ABSOLUTE SIMULTANEITY PRINCIPLE

The chapter provides an analysis of a historically influential thought experiment in physics, the Einstein's *Train Thought Experiment*, as an illustration of a successful a priori revision in natural science. The analysis is presented in four main and several sub-sections. The first main section introduces some historical background (epistemic analysis of Galileo's *Falling Bodies* thought experiment) for a specific use of the epistemic function of experience in the main argument. The second section presents John Norton's influential challenge to find a thought experiment that cannot be reconstructed as a logical argument. Norton's account has two main theses, the epistemic thesis that all information about the physical world delivered through a thought experiment comes solely from experience and the reconstruction thesis that all thought experiments could be reconstructed as arguments. I argue that at least in some cases Norton's theses are incompatible with each other and therefore, their combination could not form a reliable account. I try to show that sometimes the available experience not merely could not justify the conclusion of a thought experiment but even contradicts it. In the third section I suggest an analysis of Einstein's *Train Thought Experiment* (TTE) both as a counterexample to the challenge and as an illustration of a historically significant a priori revision in physics. In the fourth section I use the analysis of the TTE to meet Norton's challenge. In the fifth section I respond to some replies by James Robert Brown that concern the analysis of the *Train Thought Experiment*.

Galileo's Falling Bodies Thought Experiment

One of the most famous thought experiments in the history of science is Galileo's *Falling Bodies*.[1] His aim was to demonstrate by means of pure reasoning and with no help from actually conducted physical experiment that all bodies, regardless of their weight, fall at the same speed. The structure of the experiment has several distinct steps, which are connected by logical reasoning with the aid of visualization. Besides logical inference from certain premises the thought experiment makes use of what James Robert Brown and many others call "mental manipulation", i.e. imaginative suggestions, that draw the development of the thought experiment in a certain direction. One of the important and perhaps intended features of the thought experiment is its intuitive compellingness, logical certainty and visual clarity. Galileo leads us from certain well-accepted assumptions, which he connects with additional ones and after that he asks us to imagine a series of events that are consistent with them. The construction is a classic example of *reductio ad absurdum*, the contradiction reached is clear and convincing for every one who followed the reasoning.

Galileo starts with the reigning Aristotelian view about motion that heavier bodies fall faster than light ones. He ask us to imagine that a heavy cannon ball is attached to a light musket ball and urges us to examine logically what would happen if we construct mental image of them falling together in accordance with the Aristotelian requirement. Here we are presented with two consequences that follow in parallel inescapably, if we accept the Aristotelian view. On the one hand, the light ball should slow down the heavy one and so the sum speed of the combined system of falling bodies would be smaller than the speed of the heavy one when it falls alone. On the other hand, the system of combined heavy plus light ball has weight that is bigger than the weight of the alone heavy ball, so it should fall faster than the heavy ball alone. The two consequences are consistent with both the initial assumption and the suggested mental manipulation, which considered alone, is compelling as well as logically and imaginatively unproblematic. What is important, though, is that there is no way for us to eliminate any of the consequences – they both *follow* from the construction. Together, however, they form a contradiction, for their meaning is that on the Aristotelian view the heavy ball has to fall *both* faster and slower than the heavier system of the two balls. Galileo's solution is to reject the Aristotelian premise. We can briefly summarize the structure of Galileo's argument:

1. Heavier bodies fall faster than light ones (Aristotelian-Scholastic view)
2. Combined system of heavy and light body (ball) is heavier than either of them alone (intuitively certain)
3. Because of (1) the combined system's speed ($S_{combined}$) would be slowed down by the light one's (S_L) since it's speed is smaller than the speed of the heavy ball ($S_{combined} = (S_H) - (S_L)$): ($S_{combined}$) < ($S_H$)
4. Because of (1) and (2) ($S_{combined}$) should be bigger than (S_H): ($S_{combined}$) > (S_H)

5. By the law of non-contradiction a value (of a speed) could not be at the same time bigger and smaller than another, one and the same value, so we read (3) and (4) as contradiction, (2) and (5) are unproblematic, *therefore*
6. Aristotelian-Scholastic View (1) must be rejected
7. Still, we need to know which will fall faster if it falls so at all and Galileo's answer is
8. None. They both will fall with the same speed - since we have eliminated the possibility "bigger than" and "smaller than" the only choice left is "equal to"

In this reconstruction, the assumptions are (1), (2) and (5). If we set aside the logical law of non-contradiction, what we are left with is to consider the Aristotelian assumption and the intuitively certain view that combined system of bodies has bigger weight than any of them separately.

In Michael Friedman's three-fold model of scientific knowledge the Aristotelian conception of motion (1) would be classified as an empirical law. The idea that heavier bodies fall faster than light ones has some intuitive compellingness but in fact, it is justified by our everyday experience where heavier bodies *do seem* to fall faster than lighter ones. What could be the reasoning behind this? We can suggest that it is something of the following form:

> Bodies fall because of their weight. Weight seems to have the important property of causing the fall of a body. One of the most important properties of a falling body is the speed. Since the weight and its magnitude in specific seem to cause the fall of a body it must affect the properties of the fall, i.e. including its speed. It is natural to suppose that changes in weight will affect changes of speed and that increasing the weight will increase the speed. This is why heavier bodies must fall faster than light ones.

This kind of justification clearly relies on empirical premises. Experience, however, allows only for observation of non-ideal cases, i.e. cases where the air resistance, the density of the medium (air), the shape of the bodies, etc. vary. This is an indication of how powerful a tool the thought experiment could be in eliminating all the variations and providing ideal homogeneous cases where the scientist could concentrate on the intended matters under "mentally controlled" conditions. Whatever the nature of justification of (1) might be, the interesting part is that the justification of (2) and also, the epistemic reasons for *proposing* (2) in the first place, could not be experiential. Because we do not need experience, even if we previously have had an experience, in order to know that when we combine something (that could be expressed by positive value) with something else (also expressed by positive value) we will have a combination, or a set with these two as a members that has different and bigger value.

On the natural objection that (2), (5) and the steps between them are nothing but logical reasoning there is an important distinction that has to be clarified: Even that certainly the transition between the different steps is done following the rules of logic there are three things that logic alone could not do. These are:

- Logic alone could not generate (2)
- Logic alone could not propose (2) for the argument
- Logic could not hold (2) as (intuitively or a priori) true

The empiricist line tends to show that the information in the conclusion has been hidden in the premises of the argument and also that every thought experiment is expressible in argumentative form. Recently the most energetic proponent of this view is John Norton[2]. However, and this is a decisive epistemic point, the conclusion of the thought experiment could not have been epistemically justified from experience. For even if we somehow agree that experience crawls down the premises the only available experience tells us that heavier bodies *do seem* to fall faster than lighter ones. No experience is available for the opposite and thus the obvious epistemic support goes to the Aristotelian thesis which is the source of the logically detected contradiction and not to the conclusion of the *Falling Bodies*. For the relevant experience is natural experience of observation and there we have no detachment from important properties like air resistance and the like. On virtually any account of the nature of epistemic justification whatever and the epistemic source of the reasons to hold the justified proposition one thing is most clear: that the source reasons *could not undermine* what they are supposed to justify. In this sense were there reasons that justify the conclusion, which happily happens to be the right one as well, they could not have come from experience. A modal objection naturally follows. An opponent like Norton might say that even if at the time of the thought experiment such experience was not available it is nevertheless available now and in this sense it is possible to have the experience to justify the conclusion of the thought experiment. Well, not quite. For scientific epistemology cares about the epistemic nature of the reasons that do justify conclusions and not the reasons that might, in time, justify conclusions. Epistemic justification tell us about the actual reasons a scientist like Galileo is having to hold a proposition like the conclusion of the *Falling Bodies*. And inquires about the epistemic nature of *that* reasons. Clearly, any experience whatsoever that would be coming in 10, 100 or a million years, as it were, is completely irrelevant here. For Galileo has no access to such experience and thus could not build his justification upon it. A proposition, and especially in science, when invoked to deal with a given problem, has to be epistemically justified. This sort of justification illuminates us about the actual steps scientific knowledge makes. And even if it is that case that in 300 years new reasons become available this can not be helpful for they can not go back in time and substitute the actual historical justification that led to knowledge and thus contributed to scientific growth. The epistemic role of experience is thus quite clear: only experience *available* at the occasion of justification *could* justify a proposition. And no future generation or extraterrestrial sort of potential ultimate experience could ever do the job if it is not at the right place and at the right time. This line of reasoning serves to introduce the discussion where epistemic contradiction of allegedly supported proposition is revealed as a powerful weapon of the rationalist.

Reconstruction, Justification and Incompatibility in Norton's Account of Thought Experiments [3]

In a number of papers, systematically defending moderate empiricist view about the nature of thought experiments in science, John Norton famously puts forward two requirements, which, he takes, are central for the debate. The first deals with the logical reconstructability of thought experiments [TE], the second with their epistemic status. The essence of the two requirements is synthesized as follows:

> RECONSTRUCTION THESIS [RT]: All TEs can be reconstructed as arguments based on tacit or explicit assumptions. Belief in the outcome-conclusion of the TE is justified only insofar as the reconstructed argument can justify the conclusion.

The second thesis is response to *the epistemological problem of* TEs [EP] - TEs are supposed to give us information about our physical world. What could possibly be the source of this information? Norton's answer is his

> EPISTEMIC THESIS [ET]: The information can come only from experience

In this chapter, I would argue that at least some in cases Norton's RT and ET are incompatible with each other. Therefore, their combination could not form a reliable account of the actual epistemic status of TEs in science. In what follows I will try to show that sometimes the experience available could not justify the conclusion of the thought experiment and even more, it contradicts it. I would argue that the consequences of such contradiction bear seriously against both Norton's theses and my main illustration would be a single influential thought experiment from the history of physics. If successful, this would demonstrate that an influential version of an empiricist account on TE in science fails, thus weakening an important branch of the contemporary criticism against a priori rival accounts on TE, for example Brown's platonic one.[4]

Prima facie, from Norton's own examples it seems that any premise of the reconstructed argument should be justified[5] for at least two reasons: first, if it were unjustified, it would undermine the soundness of the conclusion and second, if it were unjustified it would not lead to supposedly correct knowledge about the physical world, derived empirically from true premises.[6] Thus, it is clear that the logical requirement for reconstructability of a thought experiment as a valid argument is vacuous without the epistemic requirement the argument's premises, hence the conclusion, to be justified. [7] In this sense, the logical requirement *implies* the epistemic. If we embrace the epistemic requirement for justification of the premises and the conclusion, would we be obliged to embrace the logical reconstructability requirement as well? At least it seems that this is the main line of argumentation followed by Norton. RT appears to be the main instrument which would secure the truth of the ET. But in principle there exists the theoretical possibility for a TE to be reconstructed as an argument with purely a priori justified premises and conclusion. In the face of this alternative Norton's epistemically free play with RT would look strange for an empiricist. I believe that Norton would most probably not give up the epistemic requirement, it looks vital for the logical one to pass – if the premises let alone the conclusion are not [empirically] justified the whole point of the reconstruction as an argument seems lost; only what is delivered through an argument from previously established premises would count as justification. Put briefly, both ET and RT seem bound to be together: ET with no RT is not proved and RT without ET is pointless from an empiricist's point of view. Therefore, we could reformulate the requirements as one coherent joint requirement: *epistemic reconstructability requirement* [ERR].[8] In what follows, I would try to show that the nature of such requirement concerning (some or at least one) TE in science is contradictory and that once ERR is broken down, the separate requirements for argumentative reconstruction and empirical justification fail to hold in some certain cases of thought experiment(s), which yet we have all reasons to accept as justified and successful. To illustrate this claim, I will argue that Einstein's famous Train Thought

Experiment (*TTE*) satisfies Brown's definition of a platonic TE and that it presents a prospective candidate that meets both Norton's challenges.[9]

The Formal View of the Epistemic Problem

I would like to comment briefly on two aspects on the epistemic challenge. Norton claims that there are two reasons behind his RT, namely, a) empiricism and b) that he never found a TE, which cannot be reconstructed as an argument. I find the first reason instructive about the relationship between RT and ET. When we consider a TE in science Norton's claim is that the premises of the reconstructed as argument TE are empirical. This is the epistemic part of his answer to the problem. This seems to be crucial for his general view on TE. On the one hand, it resolves the problem with the eventual *a priori* justification of the premises and hence, *a priori* justification of the conclusion; on the other hand and probably more important, provides the epistemic justificatory basis, which would be transferred through the argument to the conclusion of the TE – that is, the empirical justification.

Traditionally, an argument is considered empirically justified if its premises are empirically justified and if the conclusion follows from the premises. Borderline cases are certainly possible. Thus, for example, we might have a case of an argument where some of the premises are empirically justified and some seem to be justified independently from experience, i.e. they seem to be justified *a priori*. The question is: At the end of the day, is such an argument justified empirically or a priori? Prima facie, we could not argue for full a priori justification since the transition from the premises to the conclusion is made possible by at least one empirically justified premise. However, the opposite case appears as equally plausible – the fact that not all premises of the argument are empirically justified seems not to allow us to classify it as merely empirically justified.

These considerations seem to have direct bearing on both RT and ET, the last being translated as the claim that the premises and consequently, the conclusion of the argument, are empirically justified. For if we agree that even one premise which plays full blooded role in a TE [reconstructed as an argument] has non-empirical justification, this would lead to some immediate consequences about the RT and about its joint function with the ET to prove the empirical nature of the justification of all TE in the natural sciences. From an epistemic point of view, probably the most interesting consequence would be that the justification of the conclusion of the argument would not be entirely empirical and in the face of it, the argument's justification would not be entirely empirical. From a logical point of view, it seems that it would not be possible RT and ET to be maintained jointly since the RT would presuppose at least a single instance of non-ET premise. Norton's ERR, however, seem to rest on a supposed *justificatory homogeneity.*[10] In a scenario where there is a violation of the homogeneity, this presents problems for its scope and validity:

E_P	(empirically justified premise)
A_P	(a priori justified premise)
E_P	(empirically …)
A_P	(a priori …)
…….	
(E and/or A)$_C$	(empirically and/or a priori justified conclusion)

The force of a TE, reconstructed or not as a logical argument, lies in its relation to the physical world. A successful TE would tell us something which is true of the physical world and the unsuccessful one would fail to do so. In this sense, the epistemic value of the assumptions [premises] and the result [conclusion] of the TE plays the indispensable role of providing the epistemic basis for the relation of the claims of the TE with the physical world. They carry the load of the justification of the [resulting] claim and the epistemic nature of this justification, namely empirical or not. Any interruption of the homogeneity of the transferred epistemic values would lead to epistemically non-homogenous results. For example, if an argument with a priori justified premises is "contaminated" with an empirically justified premise it would not be homogenously a priori justified argument but it would not be homogenously empirically justified one as well; its complete and supposedly correct justification could not come from experience alone [neither from a priori reasoning alone, as it were]. The question, which seems inevitable then is:

> What is the epistemic character of the justification of an argument with epistemically diverse premises?

When we view this question in the context of the TE in science and Norton's RT and ET, it seems that what Norton is after is full empirical homogeneity of the justification of the supposed reconstructed premises and conclusion. In purely epistemic terms, this seems to be too strong a requirement. Since if a counterexample shows that a TE in natural science actually does posses even one non-empirically justified premise that would impair the epistemic homogeneity of the argument. Before to turn into this matter more closely let us see whether it is necessary for Norton to argue for epistemic homogeneity in the case of the [joint] RT and ET theses.

In terms of his reconstruction of the epistemic problem of the TE it is clear that any non-homogeneity in terms of justification of the premises would mean that he should admit of non-empirically, i.e. a priori justified premise. This directly contradicts his underlying attitude of more or less modest empiricist that "… all knowledge of our world does derive from experience".[11] Any such admittance of a priori justified premise, however, seems to carry danger for the underlying attitude in the first place; if true, it would mean that the conclusion of the argument was not justified entirely on the basis of experience and consequently, if the conclusion is correct, that there is a piece of knowledge about the physical world which is there not derived from experience but by virtue of some non-experiential consideration.[12]

Thus, I believe that the epistemic non-homogeneity is not an option for Norton's account. The task of the apriorist now is not as hard as it is usually considered to be. She does not have to find an entirely a priori justified argument with all its premises justified a priori, which could probably be found with less difficulties in mathematics than in the natural sciences. Instead, she might concede with an argument that would have at least one justified a priori premise.[13]

If the underlying attitude for RT is empiricism, then, it is natural to expect problems related to the attitude to lead to problems with the RT itself. Significant for this line of expectation is the following consideration: if there turns out to be even a single counterexample of [eventually reconstructed as a logical argument] TE with a non-empirical premise this would present a problem for Norton's account; first, ET would be false since there is a case of knowledge delivered by a TE that does not derive from experience and second, *given ET,* the point to maintain RT would be seriously weaken since it does not fulfill its role to explain [epistemically] the conclusion of the TE. In fact, Norton himself explicitly anticipates challenge of a similar kind.[14] I would try to show that a TE could be found that meets Norton's challenge. This would be possible if the TE manages to demonstrate that given *ERR*, it cannot be reconstructed as an argument. The next section deals with developing a concrete proposal for a TE that meets the challenge. My candidate is Einstein's famous *Train Thought Experiment*, which plays role for the establishing of the Special Theory of Relativity.[15] If we follow Brown's classification on TE in science, TTE seems to be a good candidate for a platonic TE. At the same time it is *destructive* – destroys the principle of absolute simultaneity, which in its own right is equivalent of destroying a physical theory, which rests on this principle, and *constructive*, since it establishes the principle of relative simultaneity, which is one of the key principles that underlie a new theory, the Special theory of Relativity.

The Train Thought Experiment and the Conflict between the Principles [16]

In this section I will take a particular case of influential scientific thought experiment from the history of science and I would argue that it could serve as an illustration for a priori revision of fundamental scientific principle(s). My thesis would be that the famous Einstein's *Train Thought Experiment* (TTE) is a paradigmatic example of an a priori revision – that it has actually been conducted a priori, starting from a priori assumptions and reaching a priori justified results. I would claim that the means through which the revision has been performed are a priori and not empirical. For that purpose I will argue that the leading *assumptions* of the thought experiment (as they appear in the thought experiment reasoning) which play the role of justificatory instruments for the conclusion(s) of the experiment are a priori justified. Also, I will argue that the relations between the assumptions lead to the revision not for some experiential considerations but purely from a priori reasons. The thesis is meant as a part of defense of the broader claim, namely, that certain thought experiments in science could serve as paradigmatic examples of a priori revision of a priori principles of science.

In brief, my argumentation would proceed in the following way. I will state the assumptions of the thought experiment, I will reconstruct Einstein's reasoning and I will try to explicate the actual epistemic justification behind both the assumptions and the conclusion. The main assumptions that take part in the thought experiment are the *Principle of Relativity (PR)* and the *Light Principle (LP;)* (as well as the implicit *Absolute simultaneity principle (AS)* which is being revised). I will argue that the Principle of Relativity is straightforwardly justified as a priori; concerning the principle of the constancy of speed of light, I will propose an analysis which attempts to show that the *de facto* role of this principle in the thought experiment is a role of an a priori justified principle.

If successful this should demonstrate the historical example of TTE as a revision of a fundamental scientific principle (the absolute simultaneity principle) conducted through a priori reasoning. This would provide an alternative to the traditional empirical revision and would defend the view about the a priori character of the principles against the traditional attack, which argues that they are not a priori but empirically revisable. The thesis for their a priori revisability defended here does not aim to disallow empirical revisability, but instead it just purports to show that a priori revision is possible and in some cases, historically actual. No matter how modest such a claim could look, I believe it bears significant epistemic consequences.

The Thought Experiment

There is no better exposition of the *Train Thought Experiment* and the revision it performs than Einstein's own original one and therefore I quote it here at length:

> "In short, let us assume that the simple law of the constancy of the velocity of light c (in vacuum) is justifiably believed by the child at school. Who would imagine that this simple law has plunged the conscientiously thoughtful physicist into the greatest intellectual difficulties? Let us consider how these difficulties arise.
>
> Of course we must refer the process of the propagation of light (and indeed every other process) to a rigid reference-body (co-ordinate system). As such a system let us again choose our embankment. We shall imagine the air above it to have been removed. If a ray of light be sent along the embankment, we see from the above that the tip of the ray will be transmitted with the velocity c relative to the embankment. Now let us suppose that our railway carriage is again traveling along the railway lines with the velocity v, and that its direction is the same as that of the ray of light, but its velocity of course much less. Let us inquire about the velocity of propagation of the ray of light relative to the carriage. It is obvious that we can here apply the consideration of the previous section, since the ray of light plays the part of the man walking along relatively to the carriage. The velocity W of the man relative to the embankment is here replaced by the velocity of light relative to the embankment. w is the required velocity of light with respect to the carriage, and we have
>
> $$w = c - v.$$
>
> The velocity of propagation of a ray of light relative to the carriage thus comes out smaller than c.
>
> But this result comes into conflict with the principle of relativity set forth in Section V. For, like every other general law of nature, the law of the transmission of light in vacuo must, according to the principle of relativity, be the same for the railway carriage as reference-body as when the rails are the body of reference. But, from our above consideration, this would appear to be impossible. If every ray of light is propagated relative to the embankment with the velocity c, then for this reason it would appear that another law of propagation of light must necessarily hold with respect to the carriage—a result contradictory to the principle of relativity.

In view of this dilemma there appears to be nothing else for it than to abandon either the principle of relativity or the simple law of the propagation of light in vacuo. Those of you who have carefully followed the preceding discussion are almost sure to expect that we should retain the principle of relativity, which appeals so convincingly to the intellect because it is so natural and simple. The law of the propagation of light in vacuo would then have to be replaced by a more complicated law conformable to the principle of relativity. The development of theoretical physics shows, however, that we cannot pursue this course. The epoch-making theoretical investigations of H. A. Lorentz on the electrodynamical and optical phenomena connected with moving bodies show that experience in this domain leads conclusively to a theory of electromagnetic phenomena, of which the law of the constancy of the velocity of light in vacuo is a necessary consequence. Prominent theoretical physicists were therefore more inclined to reject the principle of relativity, in spite of the fact that no empirical data had been found which were contradictory to this principle.

At this juncture the theory of relativity entered the arena. As a result of an analysis of the physical conceptions of time and space, it became evident that in reality there is not the least incompatibility between the principle of relativity and the law of propagation of light, and that by systematically holding fast to both these laws a logically rigid theory could be arrived at. (Einstein, Albert [1920] *Relativity The Special and the General Theory*, London, Routledge, pp. 18 – 20)"

and

"UP to now our considerations have been referred to a particular body of reference, which we have styled a "railway embankment." We suppose a very long train traveling along the rails with the constant velocity *v* and in the direction indicated in Fig. 1. People traveling in this train will with advantage use the train as a rigid reference-body (co-ordinate system); they regard all events in reference to the train. Then every event which takes place along the line also takes place at a particular point of the train. Also the definition of simultaneity can be given relative to the train in exactly the same way as with respect to the embankment. As a natural consequence, however, the following question arises:

Are two events (*e.g.* the two strokes of lightning *A* and *B*) which are simultaneous *with reference to the railway embankment* also simultaneous *relatively to the train?* We shall show directly that the answer must be in the negative. When we say that the lightning strokes *A* and *B* are simultaneous with respect to the embankment, we mean: the rays of light emitted at the places *A* and *B,* where the lightning occurs, meet each other at the mid-point *M* of the length *A* —> *B* of the embankment. But the events *A* and *B* also correspond to positions *A* and *B* on the train. Let *M'* be the mid-point of the distance *A* —> *B* on the traveling train. Just when the flashes[1] of lightning occur, this point *M′* naturally coincides with the point *M,* but it moves towards the right in the diagram with the velocity *v* of the train. If an observer sitting in the position *M'* in the train did not possess this velocity, then he would remain permanently at *M,* and the light rays emitted by the flashes of lightning *A* and *B* would reach him simultaneously, *i.e.* they would meet just where he is situated. Now in reality (considered with reference to the railway embankment) he is hastening towards the beam of light coming from *B,* whilst he is riding on ahead of the beam of light coming from *A*. Hence the observer will see the beam of light emitted from *B* earlier than he will see that emitted from *A*. Observers who take the railway train as their reference-body must therefore come to the conclusion that the lightning flash *B* took place earlier than the lightning flash *A*. We thus arrive at the important result:

Events which are simultaneous with reference to the embankment are not simultaneous with respect to the train, and *vice versa* (relativity of simultaneity). Every reference-body (co-ordinate system) has its own particular time; unless we are told the reference-body to which the statement of time refers, there is no meaning in a statement of the time of an event.

Now before the advent of the theory of relativity it had always tacitly been assumed in physics that the statement of time had an absolute significance, *i.e.* that it is independent of the state of motion of the body of reference. But we have just seen that this assumption is incompatible with the most natural definition of simultaneity; if we discard this assumption, then the conflict between the law of the propagation of light *in vacuo* and the principle of relativity (developed in Section VII) disappears." (Einstein, Albert [1920] *Relativity The Special and the General Theory*, London, Routledge, pp. 25 – 27)

[1] As judged from the embankment

Historical Exposition of the Revision

The major principles participating in the thought experiment are the Principle of relativity (in restricted sense, PR) which says that

> If, relative to K, K' is a uniformly moving co-ordinate system devoid of rotation, then the natural phenomena run their course with respect to K' according to exactly the same general laws as with respect to K for all inertial systems

the Principle of the Constancy of speed of light (LP) that

> The constant speed of light (c) in vacuo does not depend on the (magnitude and direction of) velocity of the light source

and the principle of Absolute simultaneity (AS) which, translated for the TTE case, says that

> If we consider any two frames of reference (here: the ones of the train and the embankment) events that are simultaneous with respect to one of the frames are simultaneous to the other as well. There is one common absolute time for every frame of reference. Different observers agree about simultaneity of events with respect to different frames of reference absolutely.

The structure of the revision and its steps could be identified in the following way:

1. General theoretical background (Maxwell's Equations for EM, acting mathematical theory, etc.)
2. Principle of Relativity (PR)
3. Light Principle (LP)
4. Absolute simultaneity (AS)
5. Logic reveals contradiction between PR and LP when viewed in the light of AS
6. AS critically examined through the mentally conducted TTE
7. In TTE different observers disagree about simultaneity of events (M' sees B light before A light)
8. Observer disagreement incompatible with absolute simultaneity (all simultaneous events are absolutely simultaneous for *all* observers and thus they do not disagree about them)
9. AS is exposed as causing inconsistency within the set of assumptions <PR, LP, AS>
10. AS is discarded
11. Theoretical vacuum for a principle that regulates simultaneity of events in physics
12. Either absolute or relative simultaneity, AS not an option, therefore
13. The principle of Relative simultaneity (RS) is established
14. In the light of the operational definition of RS, PR and LP are not contradictory any more: <PR, LP, RS>

What historically made Einstein discard the principle of the absolute simultaneity in the first place? The thought experiment is unambiguous about that. Among the set of assumptions (PR, LP and AS) there exists a problem. What is the epistemic nature of this problem? Is it an experiential problem? Or, it is a non-experiential one? Did Einstein detect something *in experience*, which revealed the problem? Possible empiricist's answer would be that the thought experiment *is* such experience and that Einstein just observed what happens in this, though imaginative, "experience". The thought experiment, however, by its very nature does not refer to experience to borrow new empirical data and unless the empiricist concedes that somehow the experiment generated empirical data out of pure reasoning this should be good enough reason to accept that once starting from non-experiential assumptions the *recombination* and *contradiction detection* happen epistemically independently from experience. The true nature of the problem between the assumptions becomes better visible: the problem is mere logical incompatibility, a logical contradiction between the assumptions, and thus it is not obviously of epistemic and consequently, of empirical nature. In this sense, the reason that causes the revision is nothing but the detection of logical contradiction between the assumptions. If we agree that logical contradiction is not experiential contradiction and that it is of entirely different epistemic kind we are not left with many choices. On the reading that logical contradiction is an a priori contradiction the technical reason for the revision is Einstein's detection of such contradiction. Epistemically, this shows that the reason for the revision has been independent from experience and thus an a priori one.

Besides the reason for revision there are two major components whose epistemic nature and role have to be clarified. These are the justification of the assumptions and the justification of the conclusion. With respect to this what has to be showed is that the principles in the experiment have been held as justified a priori. In order to support this line I will consider several natural objections against the a priori character of both the assumptions of the thought experiment and the justification for the transitions between them and I will respond to them. The objections can be summarized in the following three general points:

1. Motivation for the Train Thought Experiment came from empirical realm. TE is meant to explain the result of Mickelson-Morley experiment.
2. Empirical considerations tell Einstein that LP
3. Train thought experiment – assumed principles come from empirical science

The first point concerns the actual historical situation that took place with respect to the role of the Mickelson-Morley experiment. The view that major motivation behind Einstein's work on STR and on the revision of absolute simultaneity in particular was actually an attempt to account for the negative result of the Mickelson-Morley experiment, was in fact the dominant one in both popular expositions and textbooks on the subject.[17] An intense debate among historians of science and philosophers of science discussed that particular issue. The most prominent figures in the debate from the recent past are the historians of science Gerald Holton and Michael Polanyi, the psychologist Max Wertheimer and the eminent philosopher of science Adolf Grünbaum. In brief, Polanyi argued that Einstein held the principle of the constancy of the speed of light on intuitive grounds[18] and gives this as a historical example for scientific discovery as an insight into the physical reality. Wertheimer, basing on his personal conversations with Einstein in 1916 claimed that:

> "Therefore, when Einstein read about these crucial experiments …, their results were no surprise to him, although very important and decisive. They seem to confirm rather than to undermine his ideas."[19]

From this line it is clear that whatever the role of the Mickelson-Morley experiment in the discovery of STR it did not historically provide the motivation for STR. Grünbaum is the leading opponent to this interpretation, and yet he restricts himself to claiming that the evidences available on Einstein's use or lack of use of the results of Mickelson-Morley experiment are not unambiguous and are inconclusive.[20] He argues that in absence of decisive historical evidence we must accept that Einstein must have relied on the Mickelson-Morley experiment. However, the end of the debate on the historical aspect has been provided by, as Cutting puts it, the "definitive" study of Gerald Holton for the actual historical influence of the Mickelson-Morley experiment in Einstein's discovery of STR[21]. This exhaustive work showed unambiguously enough that

> "… the role of the Mickelson experiment in the genesis of Einstein's theory appears to have been so small and indirect that one may speculate that it would have made no difference to Einstein's work if the experiment had never been made at all."[22]

All this, I take, should be sufficient to show that the actual historical situation regarding the influence of the Mickelson-Morley experiment was that it had negligible significance and no motivation whatsoever for the developing STR and the revision of the absolute simultaneity in particular. As a final remark a quotation of Einstein's reply to a direct question, as described by Polanyi, is quite clear: "the … experiment had a negligible effect on the discovery of relativity". [23]

The second point is a powerful one since it claims that LP is either an empirically originated or empirically justified principle. After all, in the very fourth paragraph of the TTE exposition Einstein himself says that

> The epoch-making theoretical investigations of H. A. Lorentz on the electrodynamical and optical phenomena connected with moving bodies show that experience in this domain leads conclusively to a theory of electromagnetic phenomena, of which the law of the constancy of the velocity of light in vacuo is a necessary consequence. (*see page 180, bottom*)

Yet the question which interests us here is not the epistemic history of the Lorentzian theory, but the history of the TTE and in particular, what is Einstein's justification of the LP as a postulate in the TTE. And these two roles are pretty different for the Lorentzian theory needs ether, that carries the light (the very ether is called "luminiferous ether") whereas Einstein dismisses with ether in the beginning of his Special Relativity paper.

I will briefly comment on two aspects of this claim and then I will comment specifically on some variations of it. First, *de facto*, as Polanyi claims, Einstein held LP on "purely intuitive grounds"[24]. In the same spirit is Wertheimer's position, who traces down the origin's of Einstein's conviction in LP as far as to another thought experiment, namely the one where Einstein was puzzled how would beam of light look like if he chased it with velocity *c* (the puzzle is that in the context of the classical laws of addition of velocities the beam had to look as electromagnetic field at rest and this would violate LP)[25]. The second aspect is strictly an epistemological one. It has to meet the claim that since LP was derived from Maxwell's equations and they express a law of nature it must have been an empirical principle. The objection receives even more strength since Maxwell's equations figure in Friedman's model under the empirical part.

Much weaker variation is the one that LP has been actually established by the Mickelson-Morley experiment. The latter view is historically mistaken since the experiment has been conducted after the formulations of the equations, in the early 1880, while Maxwell's paper has been published in 1864[26]. In principle we can trace down LP to the constitutive principles that allow for its' receiving defined empirical value and thus we can argue that LP is only weakly a priori. This would allow for LP to be at least partly independent from experience even in the light of its derivability from the Maxwell's equations (ME). Such retreat, however, is not necessary. For we can argue for much stronger a priori for LP through the universality of its claim. If we take LP as formulated in the following way: *The speed of light (c) does not depend on the speed of the light source and does not depend on the direction of space for every inertial system*, it is clear that this is a physical principle with universal validity. In the case of empirical justification we are presented with two scenarios.

Either this principle has been inductively justified by some (supposedly great enough number of confirmation instances) or, it has been deductively justified following deductive rules of inference from empirical premises. The first option does not seem to take off the ground, since there are not but just a very few candidates for empirical confirmation of LP, most of which available *after* the formulation of Maxwell's equations and thus, purely historically, they could not have possibly granted the truth of the statement within its universal pretensions. The second option appeals to a possible empirical justification of Maxwell's equations. However, there is an important point about the relation that holds between the LP and the equations. In the present paper I am concerned not merely with LP but with LP *as an assumption in* the train thought experiment. This additional seemingly insignificant condition makes nevertheless a significant difference. Because, as Cutting clearly puts it

> "Einstein began with convictions that (a) Maxwell's equations are valid and that (b) Maxwell's equations – and all other laws of nature must have the same form in all inertial systems."[27]

This means unambiguously that it were not the Maxwell's equations *per se* but only the Maxwell's equations as *universally valid for all possible inertial systems* that entail the LP as an assumption in the thought experiment. Universal physical validity cannot come from inductive empirical justification. In this sense the justification of LP, whatever it is, *it could not be experiential.* In the thought experiment however, the true source of the universal validity is a universal principle that Einstein held a priori, the principle of relativity. This opens the way for a priori in a strong sense. Hence, in order to verify the epistemic status of LP as utilized in the train thought experiment, we ultimately rely on the principle of relativity, which is the substantial part of the answer to the third general point.

Now I will respond to a few short additional variations of the objections, most of them put forward by Ben-Yami in 2009 in his extensive criticism of an earlier draft of the current text. First I will discuss variations of the objection that attacks the a priori status of the LP and second, I will discuss a general objection that attacks the a priori status of the Principle of Relativity.

(1) Objection: LP is intuitive because of its relation to the Maxwell's equations and those are empirically supported by many experiments.

The LP that is related to Maxwell's Electro-dynamics (MED) is not the same principle as the one that figures in the TTE. The first one is certainly supported by many experiments but the second one is not at all. The meaning of the second LP is quite different from the meaning of standard LP that comes with the MED: the LP in Einstein's STR is *valid for all frames of reference*, none of them needs a definition with respect to a luminiferous ether and does not traverse in ether - quite an important difference. The LP in TTE is taken only together with the PR and thus, they are not just independent principles separately having a meaning on its own and independently of the meaning of the other principle, but on the opposite. In section VI Einstein is quite explicit about the actual meaning of the LP in the TTE with regards to this point.

> (2) Objection: The thesis that "universal physical validity cannot come from inductive empirical reasoning" is untenable because the Maxwell's Equations (ME) were historically inspired by many experiments.

This objection is futile. Universal validity is highly non-trivial: The light principle did not have universal validity for all inertial systems and only the Light postulate had: this is a *new* principle. Lack of the universal validity of the LP was both a part of the motivation for Einstein to formulate STR and a reason for its success. The ME were indeed supported by many experiments, but none of the experiments *supported the Light postulate*.

> (3) Objection: ME were hypothetically given universal validity because laws of physics are universally valid.

The universal validity that cannot come from inductive empirical justification is not the one of the historical ME but of the mentioned "alleged" ME, that are valid for all inertial systems. Since, there were *no* such ME, obviously no empirical support could have been available for them. In addition, since no MED were available where the Light principle were valid for *all* inertial systems there could not have been empirical support for it either – the first formulation of the Light postulate is in the STR, therefore there is no and there could not be support for it in specific *before* the formulation. At the end, the ME could not have received universal validity since they come with the ether commitment; as Einstein showed, only a new theory, with no ether commitment could accommodate the reformulated new Light Postulate. To affirm that ME are hypothetically valid does not help much with their being universally valid because of their empirical justification. But the explanation that this has been because the laws of physics are universally valid is even worse; the present text certainly accepts unconditionally the universal validity of laws. The *epistemic* question is, however, how does the empiricist know that this particular law in this particular formulation is universally valid? Merely mentioning the trivial feature of abstract physical laws (hiding the true epistemic problem of a concrete law, the thing which actually interests the text) is irrelevant and epistemically futile.

> (4) Objection: Einstein found the LP intuitive because of its relation to Maxwell's Electro-Dynamics, which were the best empirically verified theory of the day.

On the second page of the 1905 paper AE dismisses with the luminiferous ether. Both Maxwell's ED and the Lorentzian reformulation are built on the basis of the fundamental notion of the luminiferous ether, which is, to use the words of Born, "an extremely fine imponderable substance which is the carrier of the [optical] phenomena" (*Einstein's theory of relativity*, 1962, p. 86). Without the ether there would be nothing to carry the light. Since there is no ether it is quite wrong and a bit *ad hoc* to say both that that theory has been verified (for a false theory is not verified (from *veritas* – truth in Latin)) or that the light principle has been verified. Given the simple fact that Einstein dismissed with the ether and built a new theory with no ether it is clear that he did not find the principle intuitive *given* the ether. And since there is no ether and he knew that the MED was both incomplete and false (in the case of ether) it is wrong to claim that Einstein regarded the LP as intuitive. What he instead regarded was a *reformulated* version of the LP; but this was not the same principle and therefore, it could and it should not be regarded as the (or as a) principle that was empirically verified or even supported – for this principle was not available before 1905.

It is true, of course, that MED was supported by many experiments, but MED presupposed ether and no experiment supported empirically the ether claim. In fact, they all failed to show that there is an ether. And Einstein, in the first page of the 1905 paper, abandoned the idea of ether as "superfluous". The problem is that if there were ether the tension between the PR and the LP would not have arisen in the first place. So, given the trivial fact that the ether was part not only of the MED but also of the Lorentzian theory, Einstein could not have gotten his intuition about the correctness of the LP because of the empirical "support" of the experiments, that fail to show that there is ether, for the original LP was available only within an ether theory. All of this, however, is not of central relevance for the objection, for obviously the criticism does not make difference between the LP in MED and other places and the Light Postulate of Einstein – these are different propositions, with different formulation and different meaning. Therefore, no experimental support for the first could be counted as such for the second one. No experiment and no observation supported the Light postulate at the time of the STR.

A principle cannot be confirmed outside a theory in which it functions. The confirmation of the LP was only of it as a part of a theory. In fact, the very application of the LP via the PR over different and larger domain of validity is part of what makes STR such a successful and new theory. Also, the empirical verification of the LP before the Eddington expedition is very difficult to be called "verification" in the first place, for strictly it leads to no truth: there is no ether and the LP needs an ether in order to be explained. So, yes, the LP did receive great empirical support but only within the false theory of ether – that is hardly a verification. The interesting epistemic aspect of this is that the experience in the case of the standard LP has supported an incorrect theory, thus it has misled the scientists for they were interpreting the experience within a false theory. Or as Michael Friedman puts it really aptly: the principles of the theory make possible experience in the first place. The LP that has received empirical support has received such support only as a part of a theory like MED, which, unfortunately has the flaw of postulating a luminiferous ether, which, again, was rejected by Einstein in 1905. Support of a theory is not a justification of a principle that is about to enter in the theory. And especially so for a principle that not only does not figure in the theory as a proper principle of the theory but it is about to be shown (much later) that it is implied from the theory, which is a logical relation which has been shown by no one else but by Einstein.

At the third general point I will again propose a two-fold response and after that I will briefly respond to a possible objection concerning the Principle of Relativity (PR). First is the historical level response. Holton's claim is that the principle of relativity (in restricted sense) has been held by Einstein primarily on non-experimental grounds. Einstein's own words for that are "intuitively clear".[28] I take this as sufficient to enlighten the historical state of affairs regarding the actual epistemic grounds on which Einstein held PR and in particular in the thought experiment. The second line of response is a logical one: could a cognitive agent possibly hold the proposition

> If, relative to K, K' is a uniformly moving co-ordinate system devoid of rotation, then the natural phenomena run their course with respect to K' according to exactly the same general laws as with respect to K *for all inertial systems*

on grounds of experiential justification? The claim for universal validity, namely, "all inertial systems" could not be exhausted by any finite number of empirical instances of confirmation. Even few of them would not be sufficient. More, at the time the PR was accepted by Einstein there weren't any "few" confirmations for this at all. Yet, Einstein held it as true. Yet, it proved historically as a principle of immense importance for the physical science. If we accept that there is a bit of truth in PR this bit has been achieved without relying on *any* empirical justification for the truth of the principle by Einstein.

The natural specific objection that targets the a priori status of the Principle of Relativity takes somewhat of the following form:

> The principle, mentioned in the original 1905 paper by Einstein is derivable from the theory of Newtonian Mechanics and this theory had plenty of empirical support.

Such an objection, however, fails to distinguish between the *mechanical principle of relativity* (MPR) and the *general principle of relativity* (GPR). The MPR (the laws of mechanics take the same form in any of the inertial frames) does indeed have strong empirical support in the Newtonian Mechanics; whether this support is sufficient for an epistemic justification within the TTE and whether it actually did play this role is quite a separate question. But the GPR (*All laws of Physics*, that is, *electromagnetism included*, take the same form in any inertial frame) transcends in its most significant part the MPR. The transcending itself *does not come from NM* and there is neither observational nor measurement experience that *a.* justifies it and that *b.* was actually available before the STR. Therefore, we do not have *sufficient* grounds to accept the GPR is empirically justified.

The empirical support that is available is available for the Galilean Principle of Relativity. And the PR in STR is a very different principle. Einstein calls it "a conjecture" which is being "raised" to nothing else but a "postulate". The postulate is what does the real job in the STR and the same postulate figures as a principle in the TTE. The "promotion" is *not* executed in virtue of empirical reasons. To make a parallel, the same expression is used for the LP – if it were *the same* LP as the one in the MED it would not have been postulated but *it would have been accepted* as logically or empirically following from the MED. Since, however, it is not the same principle what Einstein needs, as the conjectured principle, in order to resolve a problem, it was a LP with a different meaning that was historically postulated. Now, meaning in epistemology being crucial for knowledge, I take it that any support for the old LP *is not* a support for the new LP and especially so when they are within different theories.

Therefore, even if it were the case that the old LP has actually been empirically justified this is irrelevant for the new LP. And especially so because the function of the new LP is totally new in STR and this function is only supported by Einstein's intuition or non-empirical reasoning, that is, a "reasoning that is independent from experience" and this is as good as the a priori gets. An empirical support is mentioned, but not as a reason for Einstein to accept it in the role it actually has in the STR.

A principle is, again, always a principle in a theory: outside a theory it is meaningless. The very LP is accepted because of its relation to MED. The PR from Newtonian Mechanics is for uniform motion only. The PR postulated in STR pertains to all of physics. Even if MPR has had an empirical support (and we see from the famous TE by Galileo that we can get there by pure reasoning so empirical support is not always necessary) the GPR postulate did not have empirical support at all. The extended version is thus not empirically supported and it is the extended version that participates in the TTE. The PR that was available before Einstein was relevant only for the NM and never for the electromagnetism and optics. So what we have here is principles with a very different validity and scope. It is the new principle and especially in the years *after* 1905 that the principle figures in the TTE and not the standard PR that is not valid for EM. Therefore, the specific objection fails.

The last principle assumed (though not explicitly in the initial exposition of the experiment) is the classical principle of the absolute simultaneity (AS). We may reformulate this principle using Einstein's relativistic conceptual framework as follows:

> If we consider two frames of reference (train and embankment) events that are simultaneous with respect to one of the frames are simultaneous to the other as well. There is one common absolute time for every frame of reference. Different observers agree about simultaneity of events with respect to different frames of reference absolutely.

This is the principle that is actually revised through the thought experiment. It is the fundamental principle of the classical physics and in the subsequent relativistic paradigm is substituted with the relative simultaneity principle. I will not argue in detail for the apriority of AS and technically, I do not seem to need to. For the AS principle is the principle that actually got *revised* in the TTE and thus it is clear that there was something wrong with it in the first place. In this sense its justification is of no substantial interest for it fails to deliver knowledge, due to the falsity of the principle it leads to. Nevertheless, I would like to point to an epistemically interesting detail. The principle of absolute simultaneity was (and in a standard sense it still is) quite compatible with all the available experience. For we never doubted before Einstein that events simultaneous for some observers are simultaneous for all others. In this sense, it seems that the AS principle has indeed an identifiable epistemic component in its justification which is an empirical one. This however not only does not seem to cause trouble for the a priori character of the revision of the TTE but exactly on the opposite: for if TTE conducted an a priori revision then it seems that a priori revisions could cover not only the dynamics of a priori justified principles but also of empirical ones. And this is quite welcome a result from a rationalist point of view. For it shows a greater potential for the a priori revisions. The natural reasoning here would be: if a priori revisions are capable of revising empirical principles they are quite likely able to do the same with a priori ones.

Epistemic Exposition: *TTE* as A priori Revision and Response to Norton's Challenge

In the context of *TTE* Norton's notion of "encoded" information is vague. If we follow this notion, the information exhibited in the conclusion must have been first encoded in the premises [PR, LP and PAS] and also, it must have been encoded in an empirical way. It is clear however, that no premise says something or contains information regarding the simultaneity of events except PAS and it would be really difficult to see how exactly the information about the relative simultaneity of all events is encoded or contained in the principle which asserts exactly the opposite claim. Obviously, the information is not encoded in *any* of the principles as they are by themselves. The only option left is the information to have been encoded in *the combination of the principles*. Thus, by mere logical derivation Einstein would have been able to derive the supposedly hidden there information.

The supposed encoded experience should be preserved from the premises to the conclusion of the argument. The epistemic justificatory homogeneity, presupposed by Norton, plays crucial role for the transfer and the preservation. Because the notion of the "encoding" permits somewhat broad interpretations, let us for the sake of providing a charitable reading make the claim weaker and suppose that

> The encoded in the premises experience should only be *compatible* with the conclusion.

In any case, it would be quite a puzzle how the conclusion of a supposedly empirically justified argument is incompatible with experience. Since there might be all sorts of experiences more or less relevant to the conclusion of the argument, let us restrict the experience in question just to the one that has direct bearing on the conclusion. In this sense, the claim of the conclusion should be if not directly justified by the relevant and available experience at least compatible with it.

What would qualify as experience for that case? We can distinguish between the experience in the traditional sense, like everyday observations and measurements and experience in a more elaborate sense, like real experimenting in controlled conditions. As we have seen in the case with Galileo in the first section, however, whatever kind the experience, it must have been *available* to the cognitive agent, in this case Einstein. And also, the scientific epistemologist should be able to trace the *actual* role of this experience in the building up of the justification of the conclusion of the revision. In this sense, which of the two choices (traditional and elaborate experience) could have possibly played the role in justifying (1) the premises and (2) the conclusion of the supposedly reconstructed as a logical argument *TTE*? My conjecture is that the experience in the traditional sense is in fact the only one, which could have justified the premises of the argument and not even in a direct way via convincing inductive confirmation but only via indirect way of not refuting the claims empirically. Thus, PR and LP, if justified experientially, could have been justified only through traditional experience. Why? Mainly, because elaborate experience was not available at the time; actual relativistic effects were observed later and, most importantly, after the revision was conducted. Further, the conclusion of the *TTE* makes an empirical claim not about traditional experience but about relativistic experience, which is quite elaborate according to the above classification. Still, this experience was not available. Clearly, *TTE* was not built over an inductive basis supported by instance(s) of confirmation of simultaneity events at relativistic speeds for there were no such events available to Einstein. In this sense, the only experience *available* for the epistemic justification of the conclusion was again the traditional experience. Yet, in the case of the conclusion of the *TTE*, this experience's role is not of justification but of refutation. All available traditional experience supports directly the old thesis for absolute simultaneity (AS) and thus no traditional experience was able to support the conclusion, let alone to justify it. On the opposite: the only available (traditional) experience, which could also be the sole source of Norton's information encoded in the premises, openly *contradicts* the conclusion of the *TTE*. The traditional experience tells us nothing but that the simultaneity of events is absolute. The reason, trivial in physics but important from an epistemic point of view, is that detectable relativistic effects that bear on simultaneity appear at speeds at substantial fraction of the speed of light, for example at 85% *c*; such speeds, however, were and are still not part of any traditional experience. The minuteness and negligibility of the relativistic effects at the standard speeds is the true reason why they fail to be reflected in available traditional experience. This, perhaps, is also the main reason why absolute simultaneity has been epistemically maintained for so long: no actually detected deviation of its claims has been observed at all. From an epistemic point of view the contrast between the epistemic justificatory potential of mere experience versus precise experience could hardly be greater.

Epistemically, the available experience contradicts one of the premises of the revision, PAS. Now, let us go back and see how this contradiction affects the presupposed by Norton condition of epistemic homogeneity. If Norton is correct in the case of *TTE* all premises should have been justified empirically, PAS including. Empirical justification implies empirical compatibility, i.e. if PAS is empirically justified then it is empirically compatible. As we have seen however, PAS is not empirically compatible since it is openly opposed by the available (traditional) experience. Therefore, PAS is not empirically justified. Hence, whatever the justification of PAS it is 1) not empirical and 2) not homogenous with the rest of the premises, if we suppose with Norton that they are empirically justified. Thus, *TTE* reveals epistemic problems for Norton's ERR from two separate directions. The first one shows that the available at the time experience which could be the only source of justification of premises and conclusion contradicts a premise and the conclusion. The second one violates the supposed condition of epistemic (empirical in this case) homogeneity in at least one premise (PAS) and thus exposes its epistemic character as non-homogenous with the rest of the (empirical) premises, i.e. it exposes its character as non-empirical. Both directions point on serious problems with ERR.

Could the conclusion of the *TTE* possibly be upheld through direct justification of experience? Historical and epistemic considerations tell us that no. Historically, it was not the case that LP was experientially justified as far as it figures in the *TTE*, various opinions by Polanyi,[29] Holton,[30] Wertheimer,[31] Cutting[32] and Friedman[33] support this. In addition, no experience that could have possibly confirmed PRS, the conclusion of the *TTE*, was available to Einstein at the time of the thought experiment. For it had to be a relativistic experience that involves speeds at substantial fractions of the speed of light and that, experientially and even laboratory-experientially, was historically not the case.

I tried to show that at least in one case, the only available experience could not justify the conclusion of the thought experiment. Even worse, it contradicts it. The experience, called upon in ET, in the case of *TTE* logically contradicts the conclusion of the argument. As in the case with Galileo's *Falling Bodies*, no reasonable account on the nature of epistemic justification would allow for a alleged source of justification to contradict the proposition instead of supporting it. As an effect, in the case of TTE, this effectively eliminates the possibility of RT. The consequences for Norton's both theses, RT and ET are pretty serious – lack of experiential support and presence of experiential incompatibility through violation of the epistemic homogeneity, presupposed by Norton, weaken the epistemic thesis. Since both theses are joint by virtue of their common function in Norton's account and since the main function of the thesis for the reconstruction of the TE as logical arguments is transferring kind of epistemic justification (empirical), the problems for ET pass onto RT. When there is no homogenous epistemic justification which is to be preserved via the reconstruction and when the available experience is incompatible with the conclusion of the (reconstructed as an argument) TE the very logical validity of the supposed argument fails to obtain. If all of this is correct, then, not only ERR but also the separate requirements fail to fulfill their purpose, namely, to attest the epistemic status the TE in science and to solve the epistemic problem. If available experience could not be what justifies the revision of the AS principle and at the same time we hold that there actually was a justification for it, by principle of exclusion we reach that the revision was justified a priori. The power of the empirical contradiction practically eliminates any need to argue about the separate epistemic steps in details and thus dismisses with considerations of epistemic homogeneity. Given RT and ET requirements *TTE* cannot be reconstructed as an argument.

On the Relation between Logical Reconstructability and Epistemic Justification

In his comments on the reconstruction of the *Train Thought Experiment* as a counterexample that meets Norton's challenge, Brown appropriately points out that the nature of the connection between Norton's logical reconstructability requirement and his epistemic requirement might not be completely intimate.[34] In fact, Brown argues, *the reconstruction thesis and the empirical thesis are independent*. He illustrated this by suggesting a scenario where Norton, when prompted to respond to cases from the discipline of (I take Brown here to mean pure, that is, unapplied to the physical world) mathematics takes back his requirement for empirical justification of the premises (or, ultimately, axioms). Effectively, Brown is drawing attention to a problem that seems to follow if we elaborate on the reconstruction in a more general context. The problem is that the reconstruction could not be considered complete until the more general problem of the independence between the logical reconstruction and its empirical justification is responded to. Regarding this the nature of the relation, as I see it, might turn to be a two-fold one. The one concerns the epistemology of logical reconstructions and the other concerns the role of mathematical reasoning within the practice of natural sciences, physics here would be the natural example. Besides the obvious relevance of the epistemic point these points agree quite well with the role of mathematics in Friedman's own model of scientific knowledge where they make possible the formulation of the natural laws proper. In what follows I would like to comment on both points.

The hypothetical situation where an empiricist about natural science retreats from the claim about the empirical justification of the premises of a natural scientific argument and upholds solely the reconstruction thesis in mathematical cases represents a significant step. For it shows a difference in epistemic attitude in both cases. The leading doctrine that distinguishes an empiricist in natural science is an epistemic doctrine. Thus, it is exclusively on the basis of epistemic preference that we can say the empiricist from the rationalist in science. By retreating to mathematical cases the empiricist is in fact finding a shelter from the epistemic criticism which argues that (at least in some cases) the logical reconstruction of thought experiments as arguments is *epistemically* impossible. Thus, in order to save the reconstruction thesis the empiricist might have to be ready to retreat from his epistemic claim about how we know the premises of, say, a mathematical argument. What seems striking in such a scenario is that an empiricist would prefer his logical desideratum instead of his epistemic one; I am not sure that this is an entirely obvious solution for an empiricist but the scenario looks quite interesting nevertheless. I read the independence, which Brown proposes here, as the possibility to maintain one of the theses and not the other. We can, however, distinguish between two types of independence. The first one is a unidirectional where one of theses is independent from the other. In this type either the logical thesis could be held without an appeal to the epistemic one or the epistemic thesis could be held without the logical one but it is not the case that both of them at the same time are independent of each other. The second type is a bidirectional one where both theses could be held without an appeal to the other. In the unidirectional scenario Norton could hold the logical thesis and not the empirical one; this is the case we are actually presented with. In the bidirectional scenario Norton could hold the empirical (the epistemic) thesis but not the logical one or the logical one without the epistemic one. In order for the two theses to be fully independent, however, it seems that they should be bidirectionally independent. If not, then, they are only partially independent. In the case of natural sciences the things are not immediately obvious. Certainly, Norton's agenda is to establish that thought experiments are nothing but logical arguments and in this sense the logical thesis seems to be a primary one. On the other hand, however, and especially in the case with the natural scientific thought experiment, Norton needs a story about why should we count the arguments as sound ones and not as merely valid ones;[35] for on the reading that the logical thesis is the primary one, if the reconstructed arguments were merely logically valid then the logical reconstruction thesis would be satisfied and we would not need the epistemic thesis.

Clearly, this is not good enough for Norton and for anyone facing an analysis of a natural scientific case. For a merely valid argument is completely useless in natural science. Science deals with how the world really is and logic, by itself, could not deliver. Scientists need states of affairs that correspond to what their arguments say and in this sense the only logical construction that is good for them is of course the sound argument, that is, an argument with true premises, and not only a logically valid one. For if merely logically valid arguments were good in natural science, physics could have been full of stuff like "The sun is made out of phlogiston" merely due to their being conclusions of logically valid arguments. The question about the truth of the arguments would not even arise in the scenario of mere logical validity. Certainly, this is not an option for science and therefore the reconstructed arguments should be sound arguments and not merely logically reconstructed ones. This, however, means that Norton *should* have a story about the truth of the premises of the arguments. This story is naturally his empiricism which explains how we know that the Sun is not made out of phlogiston and the like. For these reasons I take it to be the case that the logical thesis and the epistemic one are not independent in the case of natural science. Norton needs the logical claim in order to conclude that thought experiments are nothing but picturesque aids. He needs, however the epistemic claim to argue against their a priori character. Even more, he needs the epistemic claim for his logical one to receive flesh and credibility in the first place.

Therefore it does not seem that both claims are even unidirectionally independent let alone bidirectionally in natural scientific thought experiments.

My main purpose in the chapter was to show that a counterexample for Norton's challenge is available and my illustration was from the natural science. The problem of the suggested independence of the logical and the epistemic thesis does not seem to bear directly to the counterexample; I can argue that *given* Norton's theses such a counterexample is possible. In this sense, if Norton comes up with a reformulated version of his conception of thought experiments as logical arguments I should perhaps have to come up with another counterexample. I am not sure, however, that such an alternative scenario is easy to formulate. For, as I have argued above, the logical thesis is heavily dependent on the epistemic one in natural scientific cases and in this sense any formulation which argues that thought experiments could be reconstructed as logical arguments the formulation would need the epistemic thesis. Whatever the rest of the details, this seems sufficient to accept that the counterexample holds for any conception that rests on a logical reconstructability thesis. Thus, the counterexample is in force as far as there is a logical reconstructability; this was and its main intended purpose, to respond to the logical thesis. Also, I take the problem of the dependence between the logical and the epistemic thesis in the natural sciences case to be a different one from the same problem in the case of mathematics. For whatever goes in the mathematical cases science would always need a different story about the truth of its claims, be they premises in logically reconstructed arguments or not; thus, if a logical thesis is formulated about natural science it automatically needs an epistemic complement. This is perhaps the true reason for Norton to employ an epistemic requirement in the first place. Ignoring the truth-story in natural science is a clearly a no go option.

Could, however, the logical thesis and the epistemic one be independent in mathematical cases? Brown gives the case where "a picture proof of a genuine theorem can be reconstructed as a derivation from already established theorems or axioms"[36] as a scenario where Norton upholds his logical thesis. In this case the question about the relation between epistemic justification and logical reconstruction is reduced in a way to the problem of the epistemic nature of the mathematical axioms. In my view, Brown is completely right when he points that my counterexample would not endanger Norton's conception for cases like this. The *Train Thought Experiment* counterexample was devised to meet the general challenge that no thought experiment (be it natural scientific or mathematical) could be reconstructed as an argument; it was not devised to meet a specific version of it that a mathematical thought experiment could not be reconstructed as an argument. The main reason why an illustration from the natural sciences could not be used as covering mathematical cases seems to be the different epistemology. Whereas it is overwhelmingly accepted that the epistemology of natural science is one or another form of empiricism the situation in mathematics is not so accepted. In this sense the epistemic requirement, even if turns out to be interdependent with the logical one could not be used to draw a contradiction between the two because it is not clear that the acting epistemology would have problems with the logical reconstructability. It could well turn out that the epistemology goes smoothly with the logical requirement and in this case no counterexample is available and thus the epistemic requirement and the logical one could turn out to be independent in the case of mathematics. I accept this with the only remark that from an epistemic point of view it would be strange for an empiricist to build an argument against a leading a priori doctrine about thought experiments (as Brown's platonic account which is the main target of Norton) without resting on his own epistemic doctrine (empiricism). But at the end of the day it is not clear why an empiricist about natural science should automatically be an empiricist about mathematics; history shows that empiricist accounts on mathematics were never truly successful (to take Pasch about geometry and Mill about both geometry and arithmetic). Besides the bad commercial for one's own epistemic doctrine I agree that there are no other obvious reasons why one should not opt for a different sort of epistemology in the case of mathematics.

Notes

[1] Galileo, [1974] *The Two New Sciences* (Translated from the Discorsi by S. Drake,) Madison: University of Wisconsin Press, 66f.

[2] Norton, John [2004] "On Thought Experiments: Is There More to the Argument?", Proceedings of the 2002 Biennial Meeting of the PSA, *Philosophy of Science*, 71, pp. 1139-1151.

[3] Presented at the James Robert Brown Symposium in Rijeka, April, 2005. I would like to thank Nenad Miscevic, Jim Brown, Mike Bishop, James McAllister and David Davies for helpful comments and criticism. Early portion of the chaper appeared as an article in *The Croatian Journal of Philosophy*.

[4] Developed in Brown, James Robert [1991] *The Laboratory of The Mind*, London, Routledge.

[5] During the fall of 2006 John Norton confirmed in a private correspondence that it is indeed sound argument what he needs and merely valid ones would not do. I am much indebted to John for this and many other clarifications; indeed there is nothing like the live philosophical conversation.

[6] See Norton, John [1996] "Are though Experiments Just what you thought?" In *Canadian Journal of Philosophy*, Vol. 26, Number 3, Sept. 1996 and Norton, John [2004] "On Thought Experiments: Is There More to the Argument?", Proceedings of the 2002 Biennial Meeting of the PSA, *Philosophy of Science*, 71, pp. 1139-1151.
[7] Justification here is to be understood in a loose way without any commitment stronger than *reason to hold that P*, where P is any assumption of the TE or any premise of reconstructed as argument TE in Norton's scenario.
[8] Ironically, the abbreviation is instructive for the aim of the argument that follows
[9] Both ET and RT are directly pointed against James Brown's platonic interpretation of thought experiments. Brown's own view on this particular brand of TE is best expressed by his dictum that *A Platonic TE is a single TE which destroys an old or existing theory and simultaneously generates a new one; it is a priori in that it is not based on new empirical evidence nor it is merely logically derived from old data; and it is an advance in that the resulting theory is better than the predecessor theory.* In order for a TE to qualify as a candidate for a platonic TE in Brown's sense it needs to comply with his definition but in also needs to respond to the challenges put forward by Norton.
[10] For a discussion on the problem of epistemic homogeneity see the section in chapter 4
[11] Norton, John [1996] "Are though Experiments Just what you thought?" In *Canadian Journal of Philosophy*, Vol. 26, Number 3, Sept. 1996, pp 333 – 366; p. 335.
[12] The important remark that the rules of logic do not fall into the set of possible a priori justified premises has to be made here. Norton would not resist the a priority of logic and the question about the legitimacy of its rules being dressed as premises in a logical argument is not of central importance for the present purposes.
[13] The proposed epistemic analysis of TE as arguments just follows the debate proposed by Norton on his territory and attempts to show what would be the epistemic consequences for the TE within his view only.
[14] "However, more must be said, for should a thought experiment arise which does not fit the thesis, then it would constitute a refutation of empiricism by *modus tollens*. … Thus claim 1a is subject to a very stringent test. It can be refuted if anyone can find a single thought experiment which cannot be reconstructed as claimed. I am confident this will not happen. The claim in effect advances a challenge: find a thought experiment which cannot be reconstructed as an argument. To my knowledge, no such thought experiment has come to light", Norton, John [1996] "Are though Experiments Just what you thought?" In *Canadian Journal of Philosophy*, Vol. 26, Number 3, Sept. 1996, pp 333 – 366.
[15] Actually incompatibility between RT and ET could be found even in Galileo's *Falling bodies*: the only historically available experience at the time which could have been used to justify the eventual reconstruction of the TE as an argument supports the Aristotelian principle that heavier bodies tend to fall faster than light ones; this, however, is exactly the claim refuted by the TE. The experience would

contradict the TE if evoked, as Norton suggests, to justify the premises, let alone the conclusion.

[16] The reconstruction follows almost literally Einstein's text in Einstein, A. [1994] *Relativity: The Special and the General Theory*, London, Routledge.

[17] Comprehensive collection of illustrations for this is available in Holton, Gerald *Einstein, Mickelson, and the "Crucial" Experiment*, Isis 60 (1969).

[18] Polanyi, M. [1958] *Personal Knowledge*. Chicago: University of Chicago Press, pp. 10 – 11. Also see Cutting, G. [1972] *Einstein's Discovery of Special Relativity*, Philosophy of Science, Vol. 39, No. 1.

[19] Wertheimer, M. [1959] *Productive Thinking*. Enlarged edition. New York: Harper and Brothers, p. 214

[20] Grünbaum, A. [1961] *The Genesis of the Special Theory of Relativity* in "Current issues in the Philosophy of Science", Feigl, H. and Maxwell, G. (eds). New York: Holt, Rinehart, and Winston, p.45.

[21] Cutting, G. [1972] *Einstein's Discovery of Special Relativity*, Philosophy of Science, Vol. 39, No. 1, p.54

[22] Holton, Gerald [1969] *Einstein, Mickelson, and the "Crucial" Experiment*, Isis 60 (1969), p.195

[23] Polanyi, M. [1958] *Personal Knowledge*. Chicago: University of Chicago Press, pp. 10 – 11.

[24] Ibid. pp. 5 – 11.

[25] Cutting, G. [1972] *Einstein's Discovery of Special Relativity*, Philosophy of Science, Vol. 39, No. 1, p.58

[26] Gribbin. J. [2002] *Q is for Quantum*, Phoenix Press, London, p.281 for Maxwell's equations and p.288 for Mickelson's experiments.

[27] Cutting, G. [1972] *Einstein's Discovery of Special Relativity*, Philosophy of Science, Vol. 39, No. 1, p.58

[28] Einstein, A. [1959] *Autobiographical notes*, in *Albert Einstein: Philosopher Scientist*. Schilpp, P. A. (ed.) NY, Harper and Row, p. 43

[29] Polanyi, M. [1958] *Personal Knowledge*. Chicago: University of Chicago Press, pp. 10 – 11.

[30] Holton, Gerald [1969] *Einstein, Mickelson, and the "Crucial" Experiment*, Isis 60.

[31] Wertheimer, M. [1959] *Productive Thinking*. Enlarged edition. New York: Harper and Brothers, p. 214.

[32] Cutting, G. [1972] *Einstein's Discovery of Special Relativity*, Philosophy of Science, Vol. 39, No. 1. (Mar., 1972), p.58

[33] Friedman, M. [2001] *Dynamics of Reason*: *The 1999 Kant lectures*, CSLI Publications, Stanford, CA

[34] Brown's comments on the reconstruction are given in his Brown, James R [2007] *Reply to Boris Grozdanoff* in "Comments and Replies" in *The Croatian Journal of Philosophy*.

[35] Again, I am indebted to John Norton for clarifying this point in a personal conversation at the *Pittsburgh Center for Philosophy of Science* in the fall of 2006

and actually confirming that sound arguments are what he is after and not merely valid ones.

[36] Brown, James R [2007] *Reply to Boris Grozdanoff* in "Comments and Replies" in *The Croatian Journal of Philosophy*, p. 253.

CONCLUSION

The dynamics of scientific knowledge could not be spelled out sufficiently well without providing a clear account of its epistemic aspect. Inherent part of any knowledge, scientific included, is the epistemic nature of its justificatory and revisionary regulators. Therefore, every model of scientific knowledge needs the epistemic mechanism of its dynamics embedded as its structural part. Michael Friedman's model is no exception - quite on the contrary. The epistemic nature of the fundamental principles of scientific knowledge is at its core, structuring the whole second level in the model. The constitutive principles at this level provide the framework, which only allows for the formulation and testing of the empirical laws proper. Their distinguishing role is marked by their epistemic character besides their function – Friedman argues that they are independent from experience and thus a priori. His argument for their a priori nature, however, is not powerful enough to secure it. For it relies on semantic rather than epistemic considerations and thus it fails to capture the epistemic proper nature of the principles and especially the epistemic mechanism of their dynamics. For it is clear that for any proposition to be constitutive for a law or a principle is not the same property as the one of having a specific way of coming to know the very same proposition. Being constitutive is a function of propositions and says next to nothing about the actual cognitive nature of the justificatory reasons for scientists to hold them. It is quite conceivable that a proposition arrived at in a different way, say empirical observation, might be attributed this function. Thus, it is clear that even if it turns out that constitutivity bears in some remote and indirect way to epistemicity through its semantic function this would require a separate and perhaps quite complex argument on its own. Even a superficial look at the relevant literature on the relation between semantics and the notion of the a priori would demonstrate first, how controversial this relation is and second, how difficult to arrive at a commonly agreed upon solution it is. Yet in epistemology proper things do seem neither that complex nor that controversial. For there exists a more or less dominant understanding about the a priori as independent from experience and this frames common grounds for the debate between empiricists and rationalists.

The standard argument against the a priori from empirical revisability says that a proposition could not be independent from experience for its property of being empirically revisable constitutes a dependence on experience. The present text provided an epistemic alternative that allows for a proposition or a principle, as in Friedman's case, to be a priori without being endangered by empirical revisability and without being pronounced as infallible. I have suggested and developed the notion of a priori revisability that is capable of epistemically regulating the dynamics of scientific knowledge. In this way and especially in cases where alternative epistemic ways of revision are not conceivable fundamental principles of science like the second level principles in the three-layered model of scientific knowledge could receive a proper epistemic status. On the one hand they are so fundamental in the universality of their claims that chances are they are cognitively held on a priori reasons. This is their justificatory independence from experience. On the other hand, due to the sheer nature of their subject matter, they are revisable a priori and not empirically. This is their revisability independence from experience. On a more restricted reading even if it turns out that a principle is both a priori and empirically revisable, the very fact of its having the property of being a priori revisable does constitute its revisability independence from experience in a sufficiently interesting epistemic way.

To summarize, the main problems discussed in the text were the following. First, to argue for the a priori nature of at least some fundamental principles of science. These principles were taken from Friedman's influential model of scientific knowledge. Second, in order to secure their a priori character a modern and successful argument had to be constructed that is epistemic and not semantic. Third, the standard empiricist argument from empirical revisability had to be blocked in order to reach revisability independence from experience. The solution for the first problem was to provide an argument for both the justificatory and the revisability independence from experience. This argument breaks down on two lines, an epistemic one and a historical one. The epistemic one developed a modern notion of a priori founded on both epistemic justification and epistemic revisability. Together with positive argumentation for the a priori nature of relevant epistemic justifications this constitutes the epistemic foundation for the a priori argument. The historical one, built upon the epistemic one, argued that actually there are episodes in history of science where scientific knowledge *does* seem to evolve in the suggested epistemic way. The historical illustrations were specifically selected from the domain of the second level fundamental principles, one from history of geometry and one from history of physics. The revision of the famous parallel postulate of the Euclidean geometry provided the material for the former epistemic analysis and the revision of principle of absolute simultaneity provided the material for the latter epistemic analysis. The solution for the second problem was to suggest, develop and defend a modern epistemic notion, the notion of a priori revisability, which meets the most recent standards in the literature for independence from experience. The solution for the third problem was to use the notion of a priori revisability in order to block the main empiricist argument.

Some of the results from this reasoning are general and some are specific. The general concern the fields of epistemology and philosophy of science. For epistemology, a genuine alternative to empirical revisability is developed, the conception of a priori revisability. The notion provides better and more precise criteria for identifying an epistemic candidate as independent from experience. Also, it adds to the symmetry of epistemology by providing a balance to the predominant notion of empirical revisability. At the end, it provides the rationalist with a new tool to meet the powerful empiricist argument from empirical revisability. For philosophy of science the notion of a priori revisability shows its applied role not just for a priori domains like the domain of mathematics but also for the domain of natural sciences, a domain traditionally considered as exclusively empirical. In addition, a new concrete mechanism for the epistemic dynamics of science is suggested. The mechanism regulates epistemically the growth of scientific knowledge through a continuous process of constant revisions that perfects the fundamental principles of science. In the context of dynamical model of Friedman this shows that at the fundamental second level there actually is rationality that governs the evolution of scientific principles and sometimes this rationality is independent from experience. The specific results are the pragmatic defense of the a priori nature of Friedman's model as well as the two historical analyses of the scientific case studies. Also, as a specific result might be considered the response to John Norton's challenge that no thought experiment could be found that cannot be reconstructed as a logical argument. The methodology for reaching the results is not different from any traditional scientific epistemology. For the developing of the notion of a priori revisability I follow in the steps of the recent work by Albert Casullo on a priori justification. For the analysis of the *Train Thought Experiment* which constitutes the major illustration of a priori revision in physics I employ the epistemic dimension of the modern interpretation of scientific thought experiments as (sometimes) delivering a priori knowledge put forward recently by James Robert Brown. In both cases of illustrations of a priori revisions in science I draw on available resources in history of mathematics and science.

Ever since Francis Bacon human experience has played leading role in modern scientific knowledge. Modern findings of quantum mechanics and cosmology, however, show that at least in some senses this main tool of advance is limited by the very nature of its object of investigation. Therefore, before scientific epistemology stays the task of delineating the domain of trustful validity of experience with greater precision. Michael Friedman's model, which carries the philosophical heritage of epistemic rationalism and the scientific heritage of Kuhn's structure of scientific revolutions, is one of the influential attempts to take on this task. Among its best qualities are philosophical appeal and scientific adequacy. The suggested in the present text epistemic mechanism of a priori revisions contributes to the contemporary project of the philosophy of science in two ways. First, it fills what I take to be a significant epistemic gap in the model and provides a philosophically up to date account on the epistemic nature of fundamental components of the model. In addition, it incorporates epistemically the power of human rationality even in cases when uninformed by aiding experience. Second, it shows that sometimes, but suspiciously often on the verge of the switch between scientific paradigms, science evolves independently from experience. The upshot of this might be, as I see it, that the body of human science is too complex a product, regulated by a vast number of contributing factors and every disbalance, epistemic included, does not seem to reflect properly the nature of the rich scientific results. And certainly a better understanding of how science actually works would be of help for finding ways to perfect it.

BIBLIOGRAPHY

Aitchison, I.J.R. and Hey, A.J.G [2003] *Gauge Theories in Particle Physics*, Volume II: *A Practical Introduction : Non-Abelian Gauge Theories : QCD and the Electoweak Theory*, Taylor & Francis;

Aleksandrov, A., Kolmogorov, A. Lavrentiev, M. [1969] *Mathematics: Its Content, Methods, and Meaning*, MIT Press, US.

Amir Aczel [1996] *Fermat's Last Theorem: Unlocking the Secret of an Ancient Mathematical Problem*. Four Walls Eight Windows;

Aristotle, [1992] *Physics*, Books I-II, translated with introduction and notes by William Charlton, Oxford: Clarendon Press (Clarendon Aristotle Series), 13; 293 a 19.

A. Aspect, P. Grangier, and G. Roger [1982] "Experimental Realization of Einstein-Podolsky-Rosen-Bohm Gedankenexperiment: A New Violation of Bell's Inequalities" in *Physical Review Letters*, Vol. 49, Iss. 2, pp.91-94 (1982) doi:10.1103/PhysRevLett. 49.91

Audi, R. [2003] *Epistemology*, Routledge, London.

Bell, Eric T. [1961] *The Last Problem*. New-York: Simon and Schuster [1998]

Beltrami, Eugenio [1868] *Essay on the Interpretation of non-Euclidean geometry* in Beltrami, Eugenio [1835-1900] Opere matematiche di Eugenio Beltrami / pubblicate per cura della Facolta di scienze della R. Universita di Roma, Milano, U. Hoepli, 1902-1920.

Benacerraf, Paul [1973] "Mathematical Truth" in *The Journal of Philosophy*, Vol. 70, No. 19, 661-679.

Boghossian and Peacocke (eds.) [2001] *New Essays on the A priori*, OUP;

Bolyai, Janos [1831] "Appendix Scientiam Spatii Absolute Veram Exhibens" ("Appendix Explaining the Absolutely True Science of Space"), an appendix to his father's book on geometry, *Tentamen Juventutem Studiosam in Elementa Matheseos Purae Introducendi* (1832; "An Attempt to Introduce Studious Youth to the Elements of Pure Mathematics").

Bonjour, L. [1998] *In Defense of Pure Reason*, Cambridge University Press, UK.

Bonola, Roberto [2007] *Non-Euclidean Geometry*, Cosimo Classics.

Brown, James Robert [1992] "EPR as a priori science" in *The Return of the A priori*, Supplement to the Canadian Journal of Philosophy

Brown, James Robert [1992] "Why Empiricism Won't Work", PSA: Proceedings of the Biennial Meeting of the PSA, Vol. 1992, Volume Two: Symposia and Invited Papers (1992), pp. 271-279

Brown, James [1993] *The Laboratory of the Mind: Thought Experiments in the Natural Sciences*, Routledge, New York.

Brown, James R [2007] "Comments and Replies" in *The Croatian Journal of Philosophy*.

Carnap, R. [1928] *The logical Structure of the world*. Berkeley and Los Angeles: University of California Press, 1967.

—. [1934] *The Logical Syntax of Language*, London: Kegan Paul, 1937.

—. [1936] "Testability and Meaning" in Philosophy of Science, III (1936) and IV (1937)

—. [1947] *Meaning and Necessity*: a Study in Semantics and Modal Logic, Chicago : University of Chicago Press.

—. [1950] "Empiricism, Semantics and Ontology" in *Revue Internationale de Philosophie* 11, 20 – 40, Reprinted in *Meaning and Necessity*. Second edition. Chicago: University of Chicago Press, 1956.

Casullo, Albert [1988] "Revisability, Reliabilism and A priori knowledge" in *Philosophy and Phenomenological Research*, 49, pp. 187 – 213.

—. [2003] *A priori Justification*, Oxford University Press, NY.

Copernicus, Nicholas [1543] *De revolutionibus orbium coelestium* (On the Revolutions), Nuremberg, 1543 C.E.

C.L. Cowan, Jr., F. Reines, F.B. Harrison, H.W. Kruse and A.D. McGuire [1956] "Detection of the Free Neutrino: A Confirmation", in *Science* 124, 103;

Cutting, G. [1972] *Einstein's Discovery of Special Relativity*, Philosophy of Science, Vol. 39, No. 1.

De Pierris, G. [1993] "The constitutive A priori", in *Canadian Journal of Philosophy*, Supplementary volume 18, 179 – 214.

Edidin, Aron [1984] "A priori Knowledge for fallibilists" in *Philosophical Studies*, 46, pp.189-97.

Einstein, A. [1959] *Autobiographical notes*, in *Albert Einstein: Philosopher Scientist*. Schilpp, P. A. (ed.) NY, Harper and Row.

—. [1994] *Relativity: The Special and the General Theory*, London, Routledge.

Frederick Reines and Clyde L. Cowan, Jr., [1956] "The Neutrino", Nature 178, 446

Frederick Reines and Clyde L. Cowan, Jr., [1957] "Neutrino Physics" in *Physics Today* 10, no. 8, p.12.

Frege, Gottlob [1885] *Foundations of Arithmetic*, Oxford, Blackwell.

Friedman, Michael [2001] *Dynamics of Reason: The 1999 Kant lectures, CSLI Publications*, Stanford.
Friedman, Michael. [2002] *Kant, Kuhn and the Rationality of Science*, In Heidelberger, Michael and Stadler, Friedrich (eds) *New History and Philosophy of Science: New Trends and Perspectives*. Dordrecht: Kluwer.
Friedman, Michael [2010] *Discourse on a New Method: Reinvigorating the Marriage of History and Philosophy of Science*, Open Court.
—. [2013] *Kant's Construction of Nature: A Reading of the Metaphysical Foundations of Natural Science*, Cambridge University Press.
Galileo, [1974] *The Two New Sciences* (Translated from the Discorsi by S. Drake,) Madison: University of Wisconsin Press.
Gettier, Edmund L. [1963] "Is Justified True Belief Knowledge?" in *Analysis*, Vol. 23, pp. 121-23.
Gribbin. J. [2002] *Q is for Quantum*, Phoenix Press, London.
Grice, H.; Strawson, P. [1956] "In Defense of a Dogma" in *The Philosophical Review*, Vol. 65, N. 2, p. 157.
Grünbaum, A. [1961] *The Genesis of the Special Theory of Relativity* in "Current issues in the Philosophy of Science", Feigl, H. and Maxwell, G. (eds). New York: Holt, Rinehart, and Winston.
Gullberg, Jan [1997] *From the birth of numbers*, W.W. Norton and Company, NY
H. M. Edwards [1977] *Fermat's Last Theorem*. Springer-Verlag.
Hale, Bob [1987] *Abstract Objects*, Oxford: Basil Blackwell.
—. [2007] in the introduction of "The Problem of Mathematical Objects", p. 1, forthcoming. (delivered as a talk at the Central European University in November 2007).
Heath, Thomas L. (TRN), [2002] *The Works of Archimedes*, Courier Dover Publications, NY, pp. 221 – 233.
Hilbert, D. [1950] *The Foundations of Geometry*, translation by e. J. Townsend. Reprint edition, The Open Court publishing company, La Salle Illinois.
Hilbert, David, [1918] "Prinzipien der Mathematik", Lecture notes by Paul Bernays. Winter-Semester 1917-18. Unpublished typescript. Bibliothek, Mathematisches Institut, Universität Göttingen.
—. [1922] "Grundlagen der Mathematik", Vorlesung, Winter-Semester 1921-22. Lecture notes by Paul Bernays. Unpublished typescript. Bibliothek, Mathematisches Institut, Universität Göttingen.
Holton, Gerald [1969] *Einstein, Mickelson, and the "Crucial" Experiment*, Isis 60.

Kant, Immanuel [1781/87] *Critique of Pure Reason*. Trans. Norman Kemp Smith as *Immanuel Kant's Critique of Pure Reason*. London: Macmillan Co. Ltd., 1963.

Klein, Felix [1871] "Über die sogenannte Nicht-Euklidische Geometrie", in *Mathematische Annalen*, 4: 573 – 625.

Kitcher, Philip [1983] *The Nature of Mathematical Knowledge*. New York: Oxford University Press.

—. [1980] "A Priori Knowledge," *Philosophical Review* ,89: 3-23.

—. [2001] "A priori Knowledge revisited" in Bogghosian and Peackock (eds.) *New Essays on the A priori*, OUP.

Kripke, Saul [1972] "Naming and Necessity", In *Semantics of Natural Language*, edited by D. Davidson and G. Harman. Dordrecht; Boston: Reidel.

Kuhn, T. [1962] *The Structure of Scientific Revolutions*, 1st. ed., Chicago: Univ. of Chicago Press.

Lakatos [1978]. *The Methodology of Scientific Research Programmes: Philosophical Papers Volume 1*. Cambridge: Cambridge University Press.

Lambert [1766] *Theory of Parallel lines* (Die Theorie der Parallellinien) Lambert, 1895 J.H. Lambert, Theorie der Parallellinien. In: F. Engel and P. Stackel, Editors, *Die Theorie der Parallillien von Euklid bis auf Gauss*, 1895 - BG Teubner.

Lobatchevsky N.-J. [1835] "Géométrie imaginaire (M. U. Ka., 1, 3-88, 1835 ; Œuvres complètes, réimpression, 1, Kasan, 1883)".

Maddy, P. [1980] "Perception and Mathematical Intuition" in *The Philosophical Review*, Vol. 89, No. 2. (Apr., 1980), pp. 163-196.

Manfredi, Pat [2000] "The Compatibility if A priori Knowledge and Empirical Defeasibility: A Defense of a Modest A Priori" in *The Southern Journal of Philosophy*, Vol. XXXVIII, supplement, pp. 159 – 177.

Markie, Peter J. [2000] "Modest A Priori Knowledge and Justification" in *The Southern Journal of Philosophy*, Vol. XXXVIII, supplement, pp. 179 – 189.

Mill, John Stuart [1843] *A system of Logic*, London, Longmans.

Miščević, Nenad [1998] "The Rationalist and the Tortoise" in *Philosophical Studies*, Springer, Issue Volume 92, Numbers 1-2 / October, 1998 Pages 175-179

—. [2001] "Apriority and Conceptual Kinematics" in *Croatian Journal of Philosophy* (1 (1)/2001), pp. 21-48.

—. [2005] "Empirical Concepts and A Priori Truth" in *Croatian Journal of Philosophy* (2 (14)/2005), pp. 289-315.

—. [2005] *The Three Grades of Immunity*, unpublished lectures, CEU, Budapest.

Norton, John [1985] "Thought Experiments in Einstein's Work," in *Thought Experiments In Science and Philosophy*, eds. T. Horowitz, G. J. Massey, Savage, MD: Rowman and Littlefield.

—. [1996] "Are Thought Experiments Just What You Thought?" in *Canadian Journal of Philosophy*, 26, pp. 333-66.

—. [2002] "Why Thought Experiments Do Not Transcend Empiricism" pp. 44-66 in Hitchcock, Christopher (ed.) *Contemporary Debates in the Philosophy of Science*. Blackwell.

—. [2004] "On Thought Experiments: Is There More to the Argument?", Proceedings of the 2002 Biennial Meeting of the PSA, *Philosophy of Science*, 71, pp. 1139-1151.

Collected Papers of Charles Sanders Peirce, 8 vols. Edited by Charles Hartshorne, Paul Weiss, and Arthur Burks (Harvard University Press, Cambridge, Massachusetts, 1931-1958).

Parsons, Ch. [1980] "Mathematical Intuition" in *Proceedings of the Aristotelian Society*, pp.145-168

Poincare, H. [1952] *Science and hypothesis*, 1905, New York: Dover.

Polanyi, M. [1958] *Personal Knowledge*. Chicago: University of Chicago Press.

Popper, K.[1959] *The Logic of Scientific Discovery*. Hutchinson, London.

—. [1963] *Conjectures and Refutations: The Growth of Scientific Knowledge*. Routledge, London.

Proclus [1970] *A Commentary on the First Book of Euclid's Elements*, Princeton, NJ: Princeton UP.

Putnam, H. [1979] "What is mathematical Truth" in *Philosophical Papers*, Vol. I, CUP, pp. 60 - 78;

—. [1983] "Two Dogmas Revisited" in Philosophical Papers, Vol. III, CUP, pp. 87 – 97.

Quine, W.V.O. [1951] "Two Dogmas of empiricism" in *The Philosophical Review* 60 (1951): 20-43.

Reichenbach, Hans [1950] *The Philosophy of Space and Time*, Dover.

—. [1965] *Theory of Relativity and A priori Knowlegde*, Los Angeles, University of California Press.

Riemann, B. [1868] "On the hypotheses which lie at the foundation of geometry" in Ewald, William B., ed., 1996. From Kant to Hilbert: A Source Book in the Foundations of Mathematics, 2 vols. Oxford Uni. Press: 652-61.

Saccheri, G. [1733] *Euclides ab omni naevo vindicatus sive conatus geometricus quo stabiliuntur prima ipsa universae geometria principia*. Mediolani: Ex Typographia Pauli Antonii Montani. (Reprint, with facing translation by G.B. Halsted: New York, Chelsea, 1986).

Schlick, M. [1917] *Space and Time in contemporary Physics*, In Mulder and Van De Velde Schlick (1978)

—. [1918] *General Theory of Knowledge*,. La Salle: Open Court, 1985.

Singh, Simon [1998] *Fermat's Enigma*. Bantam Books;

Shilpp, P. [1919] *Albert Einstein: Philosopher-Scientist*. La Salle: Open court.

Summerfield, Donna [1991] "Modest A priori Knowledge" in *Philosophy and Phenomenological Research*, 51:39-66.

Szabo-Gendler, Tamar and Hawthorne, John (Eds.) [2002] *Conceivability and Possibility*, Clarendon Press, UK.

Tarski, A. [1944] "The semantic conception of truth", *Philosophy and Phenomenlological Research* **4**, 13-47.

Thurow, J. [2006] "Experientially Defeasible A priori justification' in *The Philosophical Quarterly*, Vol. 56, No. 225, pp. 596 – 602.

Verulam, Lord Francis [1898] *Novum Organum* or *True Suggestions for the Interpretation of Nature*, London and New York. Book II.

Wertheimer, M. [1959] *Productive Thinking*. Enlarged edition. New York: Harper and Brothers.

INDEX